The Preservation and Maintenance of Living Fungi

D. SMITH and Agnes H. S. ONIONS

COMMONWEALTH MYCOLOGICAL INSTITUTE

Commonwealth Mycological Institute
Ferry Lane,
Kew, Richmond
Surrey TW9 3AF
England.
Tel. 01-940 4086

ISBN 0 85198 524 6

The Commonwealth Mycological Institute's Culture Collection publishes regular lists of isolates available and provides a wide range of Industrial Services, including testing, identification and a Biodeterioration Centre. For further information write to the Collection at the above address.

This and other publications of the Commonwealth Agricultural Bureaux can be obtained through any major bookseller or direct from
Commonwealth Agricultural Bureaux, Central Sales, Farnham Royal, Slough, SL2 3BN, England

Printed in Great Britain by Page Bros (Norwich) Ltd,
Mile Cross Lane, Norwich, Norfolk

Foreword

Cultures of living fungi are being increasingly used in industry, including new biotechnological processes, in research, and for teaching at a wide range of levels. In addition, they are crucial in modern plant protection, in plant pathology where inoculations are routine as components of investigative work, and further as biocontrol agents in pest management schemes.

Basic mycological isolation and culture techniques are well-described in numerous manuals, but accounts of those for the preservation and maintenance of living cultures are few and rarely cover the wide range of methods now available. This text aims to rectify this situation. It describes, compares, and contrasts, methods used and tested in the Culture Collection of the Commonwealth Mycological Institute. This is one of the largest collections of fungus cultures in the world, retaining about 10000 isolates, and since 1947 has incorporated the UK National Collection of Fungus Cultures. It is also an International Depositary for Patent isolates recognized under the 1977 Budapest Treaty.

Dr Agnes H. S. Onions is Secretary of the World Federation for Culture Collections and Curator of the CMI Culture Collection; Mr David Smith is the Collection's Preservation Officer responsible for the management and improvement of the preservation and maintenance methods at CMI.

I believe that all who work with living fungi regularly, maintaining isolates for teaching purposes, as parts of on-going research programmes, or are building up in-house or national service collections, will find this manual of considerable value.

Kew
6 *October* 1983

D. L. HAWKSWORTH, *Director*
Commonwealth Mycological Institute

Contents

	Page
Foreword	iii
Introduction	vii
I Growth of cultures	1
Media	1
Temperature	2
Light	3
Aeration	3
pH	3
Water activity	4
II Preservation Techniques	5
Continuous growth	5
1 Frequent transfer	5
2 Storage under mineral oil	6
3 Soil storage	8
4 Water storage	9
Drying and preservation by induced dormancy	11
5 Silica gel storage	11
6 Freeze drying (lyophilization)	12
7 Liquid nitrogen storage	20
III Comparison of Methods of Preservation	27
Appendix 1 Formulae for media	31
Appendix 2 Growth conditions and preservation techniques of some common fungi	37
Appendix 3 Induction of sporulation in fungi by the use of near ultra-violet light	41
Appendix 4 Prevention of mite infestations	43
Appendix 5 List of suppliers	47
References	48
Index	51

Acknowledgements

The authors would like to thank the CMI staff for their advice and technical assistance with special thanks to Dr D. L. Hawksworth, Mr A. Johnston, Dr C. Booth, Dr B. C. Sutton for editorial assistancc, Mr D. Fry for photography, Miss S. A. Wade, Miss J. Griffin and Mr S. M. Ward for valuable technical assistance, and the typists Miss C. Martin, Miss D. Rogers and Mrs J. Leopard.

The authors would also like to thank the following companies for their support in the production of this manual:—
Edwards High Vacuum,
Adelphi (Tubes) Ltd.,
Imperial Chemical Industries plc., Plant Protection Division.

Introduction

The study of microfungi, even in laboratories with limited resources, usually involves the use of living cultures. These need to be kept viable at least during the experiments and if proved of importance they should be maintained alive for future work and passed to a major culture collection. Not all laboratories can afford sophisticated methods of preservation and a recent survey of the literature has shown a great interest in cheap and simple methods, particularly by plant pathologists. This booklet is a guide to methods of preservation used at the Commonwealth Mycological Institute (CMI). Simple methods are included as well as the more complex ones.

The main aim of anyone keeping a fungus culture collection is to maintain the cultures in a viable state without morphological, physiological or genetic change until they are required for future use. The ideal situation is complete viability and stability, especially for important research and industrial isolates. With the development of biotechnology and bioengineering the maintenance of this stability and viability is becoming of increasing importance. However the small teaching or research collection may have to consider additional factors such as simplicity, availability and cost.

Culture preservation techniques range from continuous growth through methods that reduce rates of metabolism to the ideal situation where metabolism is halted. It is not actually known if this situation is achieved by any storage technique but metabolism is reduced by some to such a level that it is convenient to regard it as suspended or nearly so.

One factor common to all forms of preservation is the need to start with good healthy cultures as better results are invariably obtained. Thus the best growth conditions such as the optimum temperature, humidity, aeration, illumination and media must be found and recorded. In addition a culture collection is only as good as the monitoring system to which it is subjected. In other words it is necessary for the curator to know what characters the fungus should have, either morphological or physiological, for these to be present at the outset, and for him to be able to check them or have access to facilities or personnel who can do so. Adequate records concerning the cultures should be kept.

I Growth of Cultures

For preservation of a living culture, best results are obtained when the culture is in an optimum condition of growth at the time of treatment. Although growth requirements vary from isolate to isolate, cultures of the same species and genera tend to grow best on similar media. A knowledge of the original habitat can give an indication of suitable growth conditions. Thus, isolates from jam or cake will do best on high sugar media and species occurring on leaves will probably sporulate better in the presence of light, while those from soil may grow just as well in the dark. Cultures from the tropics do better at high temperatures. Factors affecting growth are, medium or substrate, pH, water activity, temperature, light and aeration (most fungi are aerobes, so adequate aeration should be provided).

Media

Cultures are usually best grown on agar slopes in test tubes or culture bottles. A list of common media and their modifications is given in Appendix 1 and a list of common species and recommended media and growth temperatures is given in Appendix 2.

General comments on growth media

1. Most laboratories prefer not to keep a large stock of different media and the majority of isolates can be maintained on a relatively small range of media depending on the specialization of the collection e.g. medical fungi grow well on Sabouraud's medium. Some fungi deteriorate when kept on the same medium so alternatives should be available. Large diverse collections such as that at CMI keep a wide variety of media.
2. Experience has shown that cultures grow more satisfactorily on media freshly prepared in the laboratory especially natural media such as vegetable decoctions. These are usually easy and relatively cheap to prepare and can be carried out with limited facilities. Small quantities can be sterilized using a domestic pressure cooker and if necessary the pH can be adjusted using drops of hydrochloric acid or potassium hydroxide and measured using pH papers. However, proprietary media are often useful and some media for special purposes such as assay work will require very careful preparation.
3. A wide range of media are used by different workers. Authors tend to have their own favourite media, Raper & Thom (1949) use Czapek's Agar, Steep Agar and Malt Extract Agar for the growth of Penicillia and Aspergilli while Pitt (1979) in his monograph on Penicillia recommends Czapek Yeast Autolysate (CYA) and Malt Extract Agar (MEA). It is a matter of personal preference and standardisation.
4. Mucorales do well on Malt Agar (MA) and will not grow on Czapek's Agar (Cz).
5. Many fungi thrive on Potato Dextrose Agar (PDA), but this can be too rich, encouraging the growth of mycelium with ultimate loss of sporulation so a period on Potato Carrot Agar (PCA), a starvation medium, may encourage sporulation.
6. *Fusarium* species grow well on Potato Sucrose Agar (PSA).
7. Wood inhabiting fungi and dematiaceous fungi often spore better on Cornmeal Agar (CMA) and Oat Agar (OA) both of which have less easily digestible carbohydrate.
8. Cellulose destroying fungi and spoilage fungi, such as *Trichoderma, Chaetomium* and *Stachybotrys* retain their ability to produce cellulase when grown on a weak medium such as TWA or PCA with a piece of sterile filter paper, wheat straw or lupin stem placed on the agar surface.

9. All sorts of vegetable decoctions are possible and apart from the advantages of standardisation it is reasonable to use what is readily available, e.g. yam media might be preferable to potato media in the tropics.

10. *Entomophthora* species can be grown in culture on several media but are reported to do best on an egg yolk medium.

11. The introduction of pieces of tissue, such as rice, grain, leaves, wheatstraw, dung, often produces good sporulation. The use of hair for some dermatophytes has proved very successful (Al Doory, 1968).

Temperature

The majority of fungi are mesophilic, growing at temperatures within the range 10–40°C, and growing readily at room temperature (15–30°C). Others are able to grow at higher temperatures though they are still capable of growth at temperatures within the mesophilic range and are normally termed thermotolerant. Those that grow and sporulate at 45°C and fail to grow below 20°C are termed thermophilic (Cooney & Emerson, 1964). Fungi that grow at low temperature may be either psychrophilic and are unable to grow at 20°C and above or psychrotolerant and are able to grow at low temperatures and in the mesophilic range.

Table 1 Cardinal temperatures for growth of some thermotolerant and psychrotolerant fungi. (Obligate thermophiles and psychrophiles are indicated with an asterisk).

Fungus	°C	Fungus	°C
Absidia corymbifera	45*	*Malbranchea sulphurea*	45*
Acremonium alabamense	45	*Melanocarpus albomyces*	45
Acrophialophora fusisporus	45	*Melanospora funicola*	37
Aspergillus candidus	45	*Monascus purpureus*	37
A. fischeri	45	*Monographella nivalis*	5
A. fischeri var. *glaber*	37	*Mortierella wolfii*	45
A. fischeri var. *spinosus*	37	*Myceliophthora fergusii*	45
A. fumigatus	45	*M. thermophila*	45*
A. fumigatus var. *helvolus*	40	*Myriococcum albomyces*	37*
A. malignus	5	*Myrioconium thermophilum*	45*
A. nidulans	45	*Nodulisporium cylindroconium*	45
A. terreus	45	*Paecilomyces byssochlamydoides*	45
Byssochlamys fulva	37	*P. leycettanus*	45
B. nivea	37	*P. variotii*	45
B. verrucosa	45	*Penicillium avellaneum*	37
Calcarisporiella thermophila	45	*P. camembertii*	5
Chaetomium erraticum	45	*P. digitatum*	5
C. gracile	45	*P. dupontii*	37*
C. rectophilum	45	*P. funiculosum*	37
C. thermophilum	45–50*	*P. fuscum*	5
C. thermophilum var. *corophilum*	45–50*	*P. lanosum*	5
C. thermophilum var. *dissitum*	45*	*P. lividum*	5
Cladosporium herbarum	10	*P. piceum*	37
Coprinus cinereus	40	*P. rotundum*	37
C. lagopus	40	*P. stipitatum*	37
Corynascus thermophilus	45*	*P. vermiculatum*	37
Chrysosporium thermophilum	45*	*Phanerochaete chrysosporium*	40
Dactylaria gallopara	45	*Rhizomucor pusillus*	35–45
Dactylomyces thermophilus	45*	*R. miehei*	40–45
Eupenicillium baarnense	5	*Rhizopus arrhizus*	RmT–37
E. brefeldianum	37	*R. microsporus*	45
E. egyptiacum	5	*R. oligosporus*	45
E. shearii	37	*R. rhizopodiformis*	45
Fusarium psychrophilum	5	*Scytalidium thermophilum*	45*
F. solani	5	*Sporotrichum pulverulentum*	45
Geniculodendron pyriforme	5	*Stilbella thermophila*	45*
Geosmithia argillacea	45*	*Thermoascus aurantiacus*	45*
G. cylindrospora	37	*T. crustaceus*	45*
G. emersonii	45	*Thermomucor indicae-seudaticae*	37*
G. swiftii	37	*Thermomyces ibadanensis*	45*
Gliomastix cerealis	5	*T. lanuginosus*	45
Hamigera striata	45	*T. stellatus*	45
Humicola grisea var. *thermoidea*	45*	*Thielavia australiensis*	45
H. insolens	45	*T. heterothallicus*	45
Hypocrea psychrophila	5*	*T. sepedonium*	37

This list includes fungi and incubation temperatures at which they can grow but only those marked with an asterisk show a requirement for temperatures near to those stated for growth to occur and do not grow well at room temperatures.

A list of some of these fungi is given (Table 1). Some produce different sporing states according to temperature e.g. *Thamnidium, Penicillium gladioli, Aspergillus.*

Light

Although some cultures grow quite well in the normal incubator, which is usually dark, many seem to grow better in the daylight and some spore better under black light (near U.V.) (Appendix 3). At the CMI most cultures are grown in daylight. (Appendix 2). Some are placed under black light and a few in direct sunlight. Most leaf and stem fungi are particularly light sensitive.

Aeration

Fungi are aerobic, and when grown in tubes or bottles normally obtain sufficient oxygen through the cotton wool plug, or a loose bottle cap. (Care should be taken to see that culture lids are not accidentally screwed down tightly.) However, a few water moulds require aeration by bubbling.

Water Activity a/w

All microorganisms require water for growth, but the available water needed varies according to species and isolate. The majority require quite high quantities of available water but a few organisms are able to grow at low a/w. Those which grow on jam or saltfish are sometimes quite sensitive to the amount of available water and will only grow well on media containing high concentrations of sugar or salt. These fungi may be referred to as xerophiles, osmophiles or halophiles. (Table 2). An osmophile is an organism capable of growth in environments of high osmotic pressure and halophiles require the presence of salt (sodium chloride) in the medium for growth to occur.

Table 2 Fungi that are capable of growth at low water activity.

	a_w	T		a_w	T
Alternaria citri	.84	Opt.	*A. wentii*	.84	25
Aspergillus amstellodami	.70	25	*Chrysosporium fastidium*	.69	25
A. candidus	.75	25	*C. xerophilum*	.71	25
A. carnoyi	.74	25	*Clavariopsis bulbosa*	NM	Opt.
A. chevaleri	.71	33	*Eremascus albus*	.70	25
A. conicus	.70	22	*E. fertilis*	.77	25
A. cremeus	NM	Opt.	*Monascus bisporus*	.61	25
A. echinulatus	.62	Opt.	*Paecilomyces variotti*	.84	25
A. flavus	.78	33	*Penicillium brevicompactum*	.81	23
A. fumigatus	.82	Opt.	*P. chrysogenum*	.79	25
A. glaucus (group)	.73	Opt.	*P. citrinum*	.80	25
A. halophilicus	.68	Opt.	*P. citreoviride*	.80–85	Opt.
A. medius	NM	Opt.	*P. cyclopium*	.81	25
A. nidulans	.78	37	*P. expansum*	.83	23
A. niger	.77	35	*P. fellutanum*	.80	25
A. ochraceus	.77	25	*P. frequentans*	.81	23
A. repens	.74	21	*P. islandicum*	.83	31
A. restrictus	.75	25	*P. martensii*	.79	23
A. ruber	.71	21	*P. palitans*	.83	23
A. sejunctus	.70	25	*P. griseofulvum*	.81	23
A. sydowii	.78	25	*P. puberulum*	.81	Opt.
A. tarmarii	.78	33	*P. spinulosum*	.80	Opt.
A. terreus	.78	37	*P. viridicatum*	.80	Opt.
A. tonophilus	NM	Opt.	*Sporendonema sp.*	.81	Opt.
A. versicolor	.78	37	*Wallemia sebi*	.70	Opt.

Opt., optimum growth temperature
NM, not measured, the fungus will grow on high sugar content media with low water activity
The temperature (T) is that at which the water activity was measured.
This list of organisms includes fungi that are able to grow at reduced levels of water activity. They may not all require these conditions for growth but the conditions may be preferred or even used as mechanisms of selection preventing some possible contaminants from growing.

pH

Fungi have variable pH requirements but almost always grow best in acid conditions, normally pH 5–6. Some will grow in very acid environments, pH 2 and below; these include *Moniliella acetoabutans* in pickles and *Aspergillus niger* used in the production of citric acid.

II Preservation Techniques

Good healthy cultures can be maintained for years by growth on suitable media under favourable conditions providing they are transferred to fresh media before the nutrients are exhausted and the culture stales. This requires constant vigilance of a knowledgeable operator to check that the subcultures are taken at the correct time and from typical parts of the culture and that no infections or mutations supplement the original organism. The cultures are at risk each time they are handled and transfer is time consuming so various expedients to increase the period between transfers are employed.

Continuous Growth

1. Frequent transfer

Cultures are transferred from one medium to another as required using the most suitable growth conditions. On reaching optimum growth or sporulation they are stored until reculturing is necessary. For storage purposes cultures are prepared on agar slants in culture bottles or tubes. The receptacle is a matter of personal choice. Flat bottomed bottles have the advantage that they are stable and facilitate storage. Some have screw caps which allow long storage without drying out of agar prior to use. However, once opened and inoculated, caps must be left loose to allow passage of air though this may allow the access of mites (Appendix 4). Some workers prefer tubes plugged with cotton wool, but preparation of the plugs is labour intensive. Recently some screw caps have been introduced which allow the free diffusion of gases but prevent the access of mites (Smith, 1978).

Method used at CMI

At the CMI 1 oz Universal Glass containers with plastic caps (Adelphi (Tubes) Manufacturing Ltd., 5) containing an agar slope are usually used. The reference number is written on the cap for easy identification. Plastic containers (Sterilin, 7) are used if black light is necessary to induce sporulation.

Storage

Cultures are stored in racks (Denley Instruments Ltd., 14).

(A) *At room temperature.* For short term storage; the survival and therefore subculturing period varies according to the sensitivity of the isolate. Some water moulds and medical fungi have to be cultured every 1–2 weeks, while some Penicillia and Aspergilli survive for years. However, subculturing can usually be at 2–6 months.

(B) *In cool or cold storage.* The periods between subculturing can be extended by storage at cooler temperatures, which slows the rate of growth and also the rate of variation.

(i) Storage at 4–7°C in a refrigerator or cold room can extend the transfer interval to 4–8 months. However, (a) condensation in compact collections can cause problems by enhancing bacterial or cross contamination; and (b) the humidity may rise in tightly packed collections to a point where cold resistant moulds can grow on the outside of the containers and cause contamination.

(ii) Storage in an air conditioned room at 15°C and 60% relative humidity gives adequate survival (von Arx & Schipper, 1978).

(iii) Storage of cultures in a deep freeze (−17°C to −24°C) for 4–5 years can be achieved with the majority of fungi (Carmichael, 1962). However, in the event of freezer failure, the lid should be lifted after the first few hours because overheating may occur as the cultures begin to regrow.

The disadvantages of frequent transfer

(i) Danger of variation, loss of pathogenicity or other physiological or morphological characters.

(ii) Danger of manual selection from non-typical parts of the culture.

(iii) Danger of contamination by airborne spores or mite carried infections.

(iv) It requires constant specialist supervision to ensure that the fungus is not replaced by a contaminant or subcultures made from debris or a changed sector.

(v) It is labour intensive and time consuming.

Advantages of frequent transfer

(i) Collections can be kept stable for many years provided they are supervised by a specialist.

(ii) The method is cheap and requires no specialised equipment and for a small collection the time involved is not great.

(iii) Retrieval is very easy.

2. Storage under mineral oil

Covering cultures on agar slants with mineral oil prevents dehydration and slows metabolic activity and growth through reduced oxygen tension. This method was first used extensively by Buell & Weston (1947), and subsequent reports have indicated its wide application and success (Dade, 1960; Smith & Onions, 1983).

Method used at CMI

(i) Mature healthy cultures are covered by 1 cm of mineral oil (Liquid Paraffin or Medical Paraffin specific gravity 0.830–0.890 g). The oil is difficult to sterilise but heating to 121°C for 15 minutes on two occasions seems to be satisfactory. Short slants require less oil to cover them and reduce the difference in depth of oil which occurs at the top and bottom of the agar slope. Coverage must be complete as strands of mycelium left exposed may act as wicks to dry out the culture. The oil is best prepared and added as single doses because dry spores on a culture surface may be dislodged at the time of addition and infect the oil applicator if it is used for multiple additions.

(ii) Storage is at room temperature or lower (15°C) if possible (Plate 1).

(iii) Retrieval is by removing a small piece of the fungal colony with a culture needle, hook or loop, draining off as much oil as possible and streaking the inoculum onto agar in plates or tubes. Tilting the plate or bottle may facilitate drainage. The first subculture often has a reduced growth rate and appears slimy and a second transfer is usually required before a good culture is obtained.

(iv) Intervals between transfer depend upon the sensitivity of the isolate to be stored

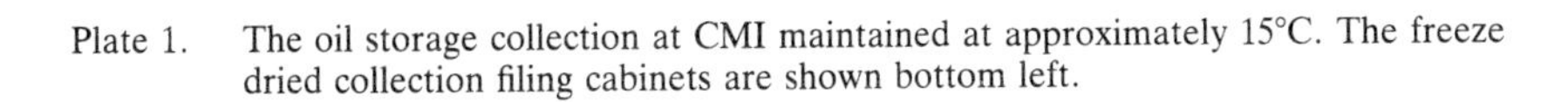
Plate 1. The oil storage collection at CMI maintained at approximately 15°C. The freeze dried collection filing cabinets are shown bottom left.

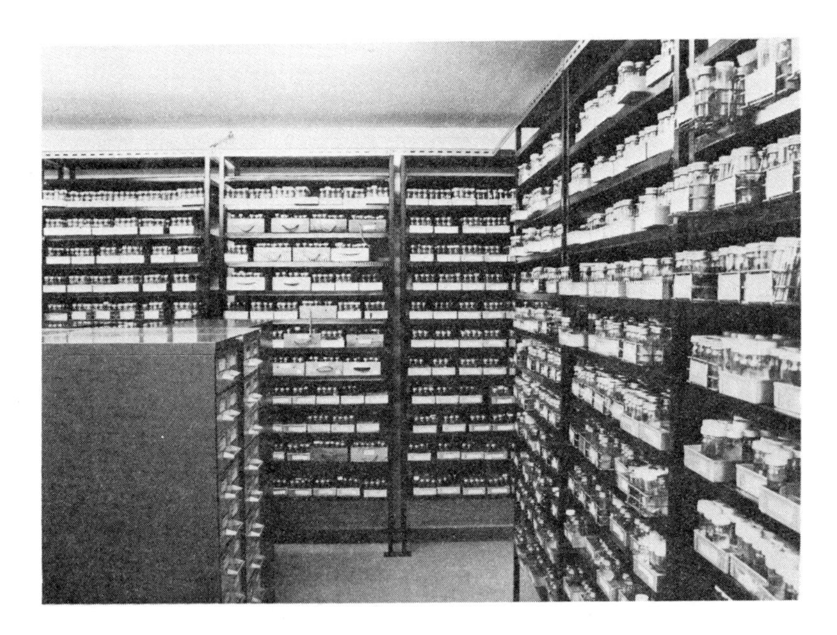

such as 1 year for water moulds. If the technique is the only available method of storage, transfers should be fairly frequent such as every two years for *Phytophthora, Pythium* (Table 3), while others such as the Aspergilli and Penicillia can be left for 30 years or more (Smith & Onions, 1983).

A wide range of fungi survive this method. Saprolegniaceae and other water moulds survive 12–30 months (Reischer 1949), whereas storage of only 1–6 weeks is normally achieved for these cultures when maintained growing on agar or in water. Other cultures have survived much longer periods at the CMI, for example revivals after 32 years for *Penicillium* and *Aspergillus* (Smith & Onions, 1983). This method is also used at the CMI for long term storage of sensitive isolates, those lacking spores or that do not freeze dry satisfactorily, such as *Cercospora, Arthrobotrys, Colletotrichum, Conidiobolus, Corticium, Nodulisporium, Phytophthora, Pythium,* and mycelial Basidiomycetes.

Table 3 Genera stored under a layer of mineral oil requiring regular transfer every two years.

Acremoniella	*Helicosporina*
Alternaria	*Heliscus*
Arthrobotrys	*Hypomyces*
Basidiobolus	*Khuskia*
Beltrania	*Lophiostroma*
Beltraniella	*Mesobotrys*
Calonectria	*Melanospora*
Cercospora	*Monacrosporium*
Chaetosphaeria	*Monascus*
Chalaropsis	*Mortierella*
Chrysosporium	*Mycovellosiella*
Claviceps	*Nectria*
Colletotrichum	*Nodulisporium*
Cylindrocarpon	*Penicillifer*
Cytospora	*Periconia*
Echinosporangium	*Pseudocercospora*
Endophragmia	*Pyrenochaeta*
Georgefischeria	*Pyrenophora*
Gloeosporium	*Spermospora*
Glomerella	*Stigmina*

Disadvantages of oil storage

(i) Contamination by airborne spores (advisable to keep a stock culture for use if contamination occurs).
(ii) Retarded growth on retrieval.
(iii) Continuous growth under specialised conditions may prove highly selective, e.g. *Fusarium* spp. rapidly become mycelial. Some ascomycete species lose the ability to produce ascospores.
(iv) Oil may splutter when sterilising the culture needle so producing an aerosol which can be highly dangerous when handling pathogens.

Advantages of oil storage

(i) Long viabilities of some specimens.
(ii) Survival of species which will not survive other methods of maintenance.
(iii) Cheap and does not require expensive equipment.
(iv) Mites do not penetrate oil cultures.

3. Soil storage

Spore suspensions in sterile water added to garden loam, in a culture bottle, allowed to grow for a few days and then left to dry while stored in a refrigerator remain viable for long periods and also can remain remarkably stable. There is an initial growth period while the fungus uses the available moisture before dormancy. This may be sufficient in some cases to allow selective growth of a mutant strain (Booth, 1971). Atkinson (1953) has obtained good viabilities of *Rhizopus, Alternaria, Aspergillus, Circinella* and *Penicillium*. The method has also proved successful with *Fusarium* spp. (Gordon, 1952; Booth, 1971) and is used at CMI for storage of this genus (Table 4).

Table 4 The isolates in the soil collection surviving up to 15 years storage.

Genus	Number of Species	Number of Isolates
Calonectria	6	8
Cylindrocarpon	12	32
Cylindrocladium	5	9
Fusarium	55	652
Gibberella	4	6
Melanospora	2	9
Nectria	17	47
Thielavia	1	1
Totals	102	764

No loss in pathogenicity of *Septoria* species isolated from cereals was observed after 20 months storage in soil (Shearer et al, 1974) or of *Pseudocercosporella* for 1 year in soil (Reinecke & Fokkema, 1979). However, examination of Gordon's collection showed that 76% of *Fusarium equiseti,* 75% of *F. semitectum* and 50% of *F. acuminatum* isolates had been outgrown by mutant strains (Booth, 1971). Despite this, soil storage should be used in preference to oil storage for the preservation of *Fusarium* species and other fungi that show variation under oil.

Plate 2. Storage in sterile soil or sand. From left to right; the culture in a plastic universal bottle, spore suspension in distilled water, distilled water, sterile soil, sterile sand, preserved fungus normally stored at +4°C but can be kept in the laboratory.

Method used at the CMI (Plate 2)

(i) Garden loam (20% moisture content) is placed in a 1 oz universal bottle to about $\frac{1}{3}$ full, and the bottle autoclaved twice (121°C for 15 minutes) 24 hours apart.
(ii) One millilitre of spore suspension in sterile water is added.
(iii) The soil cultures are left to grow at room temperature for 10 days (2–3 days for *Fusarium* spp., and other fungi with rapid growth rates).
(iv) The cultures are stored with the bottle caps loose in a refrigerator at 4–7°C.
(v) Retrieval is by sprinkling a few grains of soil containing the fungus on to a suitable medium, and incubation in optimum growth conditions.

Advantages of storage in soil

(i) Although some variation has been recorded, the subcultures have generally been remarkably stable.
(ii) Survival is good up to 10 years.
(iii) Repeated inocula can be obtained from the same sample, though it is advisable to keep a stock culture for use only if contamination or other problems occur.
(iv) The cultures are unlikely to be infested by mites.
(v) The method is cheap, does not require expensive equipment and is not labour intensive.

4. Water storage

A simple and inexpensive method of culture preservation was described by Castellani (1939, 1967) in which he stored medical fungi in water. Figueiredo (1967) and Figueiredo & Pimentel (1975) successfully maintained viability and pathogenicity in fungal plant

(vi) Caps are screwed down and the bottles stored at 4°C (though storage at room temperature is quite satisfactory) in air tight containers over indicator silica gel to absorb moisture.

(vi) Retrieval is by scattering a few crystals on a suitable medium.

A wide range of sporulating fungi survive this method (Perkins, 1962; Ogata, 1962; Onions, 1977; Smith & Onions, 1983). Thin walled spores, spores with appendages and mycelial cultures tend not to survive. Success often depends upon healthy sporulating cultures and it can be shown to differ from isolate to isolate. The results of storage at CMI are given in taxonomic form as detailed lists are impracticable (Table 6). The technique has proven to be a medium term storage method that can be used when freeze drying is not available, though the range of fungi surviving and longevities are not as good (Table 7).

Advantages of silica gel storage

(i) It is cheap and simple, not requiring expensive apparatus (Plate 3).
(ii) It produces very stable cultures.

Table 7 A list of genera surviving over 10 years storage in silica gel.

Alternaria	*Myceliophthora*
Ascobolus	*Paecilomyces*
Aspergillus	*Penicillium*
Aureobasidium	*Pestalozziella*
Beltraniella	*Peziza*
Botryotrichum	*Phoma*
Byssochlamys	*Phycomyces*
Chaetomium	*Piptocephalis*
Colletotrichum	*Polystictus*
Coprinus	*Rhizopus*
Curvularia	*Saccharomyces*
Endothia	*Schizophyllum*
Eremascus	*Sordaria*
Fusarium	*Sporormiella*
Gelasinospora	*Stachybotrys*
Gliocladium	*Thermoascus*
Humicola	*Thielavia*
Isaria	*Verticillium*
Mucor	

(iii) Penetration by mites is prevented as mites cannot survive these dry conditions.
(iv) Repeated inocula can be obtained from one bottle, though it is recommended that a stock bottle is kept in case contamination occurs during retrieval.

Disadvantages of silica gel storage

(i) It is limited to sporulating fungi, and is unsuitable for *Pythium, Phytophthora* and other Oomycetes, mycelial fungi or fungi with delicate or complex spores.
(ii) Possibility of introducing contaminants by repeated retrievals.

6. Freeze drying (lyophilization)

Lyophilization, the preservation of fungi by drying under vacuum from the frozen state by sublimation of ice, has been in regular use since Raper & Alexander (1945) reported good survivals for fungi at the National Regional Research Laboratory (NRRL), Peoria,

Plate 3. Silica gel storage. From left to right; culture, in plastic universal bottle, which has been stimulated to sporulate using near ultra violet light, spore suspension in 5% skimmed milk, sterile 6–22 mesh silica gel crystals, the preserved organism which is stored at 4°C over indicator silica gel in air tight containers.

USA. Reports on these same cultures appear from time to time indicating ever lengthening periods of survival (Ellis & Roberson, 1968) and no doubt many are still viable after 35 years.

Basic principles

(i) In general the cultures to be freeze dried should be healthy and sporulating well to achieve the best results. Very few isolates are improved by preservation.

(ii) The suspending medium is chosen to give protection during the process and also for convenience, enabling easy filling of the ampoules or vials. The media should protect the spores from freezing damage, and hopefully storage problems such as oxidation. Those normally used are skimmed milk, serum, peptone, various sugars or mixtures of them. Some sugars have associated problems due to their behaviour during freeze drying. Glucose solutions and nutrient broth, for example, can glaze and the skin formed can prevent proper drying. Bubbling can also occur prior to freezing or later in the process if the solutions are allowed to thaw. This gives reduced viabilities after relatively short storage periods.

(iii) The rate of freezing is a very important factor that cannot be ignored. Slow rates of freezing have been successful with fungi and 1°C/min is the rate normally quoted as best (Hwang, 1966; Heckly, 1978). Centrifugal freeze drying (Plate 4) normally employs evaporative freezing under vacuum which gives a favourable rate of cooling but the mechanism may be damaging in itself. More rapid cooling by immersion in freezing mixtures may give poorer viabilities, and perhaps it would be better to employ a controllable freezing stage giving the best freezing rate for each organism.

(iv) Overdrying will kill or in other cases cause mutation by damaging DNA (Ashwood-Smith & Grant, 1976). A residual moisture content of between 1 and 2% proves to be successful at CMI (Smith, 1983a).

Plate 4. The Edwards 5P5 centrifugal freeze drying machine with the centrifugal head attached and the secondary drying manifold detached standing on the top left of the machine.

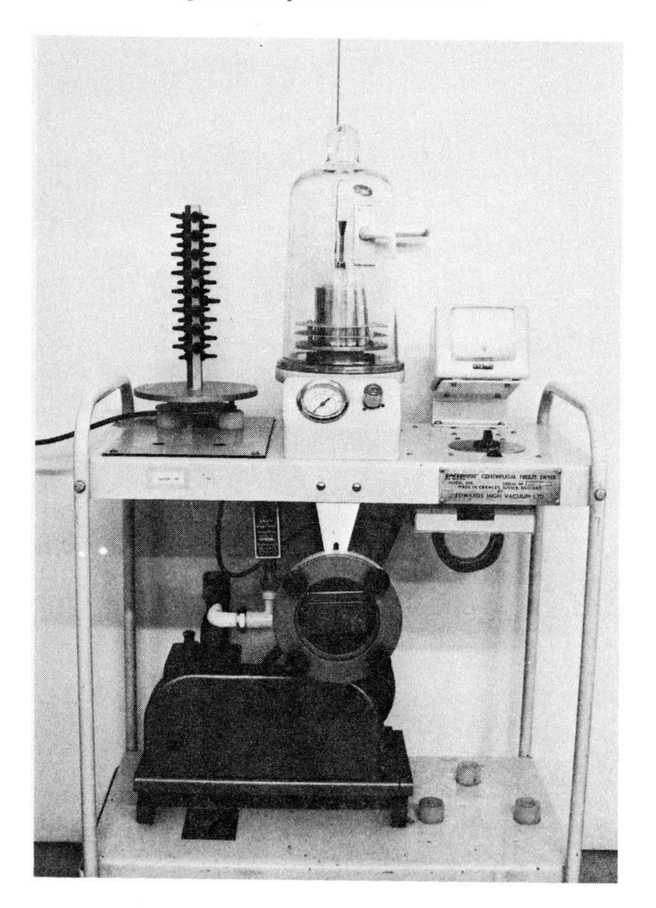

(v) Storage of the freeze dried material should exclude the presence of both oxygen and water vapour (as they can cause rapid deterioration) (Rey, 1977). This can be achieved by filling the ampoules or vials with a dry inert gas, such as nitrogen or argon, or by storage under vacuum, with a good seal (Plate 5).

(vi) Storing the ampoules at low temperature is thought to give greater longevities and 4°C seems to be favoured (Heckly, 1978). At CMI the ampoules are stored at between 15 and 20°C and fungi have survived over 15 years (Smith, 1983a).

(vii) Rehydration of the fungi should be carried out slowly giving time for absorption of moisture before plating on a suitable medium. It is possible that rehydration should be carried out in a controlled environment for specially sensitive strains (Staffeldt & Sharp, 1954).

(viii) It is advisable to rehydrate and check viabilities on a regular basis. At CMI checking and reprocessing are carried out after 10 years storage.

Methods of freeze drying

The expense of the often quite complex modern equipment can be a deterrent. However, it is not essential to have expensive equipment to carry out freeze drying

Plate 5. A selection of freeze dried ampoules and vials. Those used at CMI are the 2 ml rubber bunged vials (top left) and the 0.5 ml neutral glass amouple (shown horizontally below the vials). A freeze dried ampoule of Fleming's strain *Penicillium* is shown at the bottom of the photograph. The double ampoule shown on the extreme right is one produced by the American Type Culture Collection. It contains the freeze dried material in an inner tube and a desiccant at the base of the outer vial. (Alexander et al., 1980).

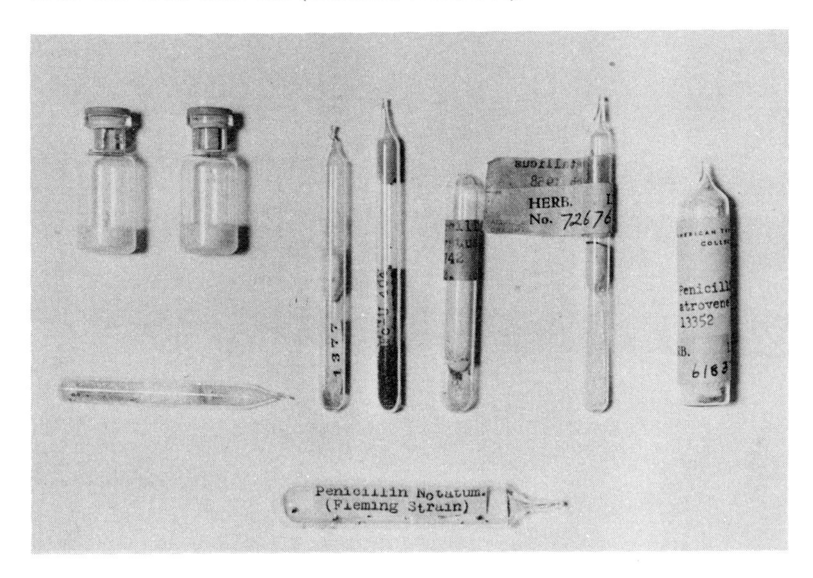

successfully. The basic requirements are a means of freezing the suspension, of generating and maintaining a vacuum, and of absorbing the water vapour evolved (Fig. 1). The ampoules containing the spore suspension can be attached to a vacuum pump via a desiccant trap and immersed in an alcoholic freezing mixture or Cellosolve (methyl or ethyl alcohol + solid CO_2 + ice) to freeze the culture, and then the ampoules evacuated. Drying is more rapid if the ampoules are removed from the freezing mixture after their contents have frozen but care must be taken not to allow thawing during drying. The ampoules are sealed under vacuum. Such an apparatus was used by Raper & Alexander (1945). A vacuum gauge is useful to monitor the process giving an indication of freezing and the degree of dryness. A non-return valve between the

Fig. 1 The basic requirements for freeze drying

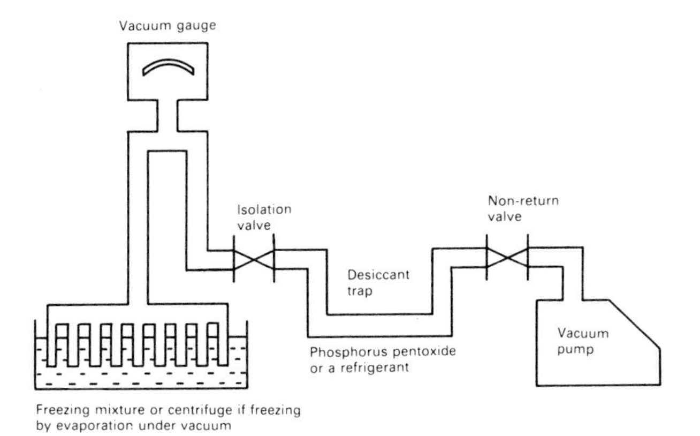

vacuum pump and the system will prevent sucking back of vacuum oil in the event of electrical failure. Using a suitable suspending medium aids the achievement of optimum residual moisture. The use of skimmed milk protects the culture during freezing and the sugars present help to retain some moisture. A recommended final moisture content is between 1 and 2%; this can be measured by oven drying or by other simple methods (Baker, 1955). When using such simple apparatus careful investigation by trial and error is necessary to achieve the best results.

At the CMI a centrifugal two stage process is in use in which freezing is by cooling by evaporation. This gives a low freezing rate and good percentage revivals are obtained. Various suspending media can be used including mixtures of skimmed milk and inositol, and sucrose and peptone.

Method used at CMI

(i) Batches of fiteen 0.5 ml neutral glass ampoules are labelled with the IMI number and current date using a reverse type printer (Rejafix, 1). These are then heat-sterilized at 180°C for 3 hours which also fixes the ink.

(ii) A spore suspension is prepared in 10% skimmed milk and 5% inositol (sterilized for 10 minutes at 115°C—over-sterilization may caramelise the milk). 0.2 ml quantities are pipetted into the ampoules and covered with a lint cap in the filtered air of a clean air cabinet.

(iii) The ampoules are placed and balanced in a centrifugal container of the EF6 primary drier (Edwards High Vacuum, 2) (Plate 6) and lowered into the chamber.

(iv) They are centrifuged, evacuated, and dried for 3 hours. The centrifuge is switched off when the spore suspension has frozen in a slant. The chamber walls are heated to accelerate the drying, and the water vapour evolved is absorbed in a refrigerated condenser.

(v) The chamber is returned to atmospheric pressure. The ampoules are removed, plugged with sterile cotton wool and constricted above the plugs using an air/gas torch

Plate 6. The Edwards EF6 centrifugal freeze drying machine with the centrifugal carriers on the top left and the Rejafix ampoule printer to the top right.

(Buck & Hickman, 3). This process should be done as rapidly as possible because exposure to air involves risk of deterioration of the partially dried suspension (Rey, 1977).

(vi) The ampoules are then attached to the Edwards high vacuum 30S2, secondary drying machine, evacuated and dried for about 17 hours, leaving a residual moisture of 1–2%. The evolved water vapour is absorbed by phosphorus pentoxide. The ampoules are sealed using a crossfire burner (Adelphi (Tubes) Manufacturing Ltd., 5; T.W. Wingent, 4) under vacuum.

(vii) An ampoule is opened after 3 or 4 days storage, reconstituted with sterile distilled water and streaked on to an agar plate to test viability. The water is allowed to absorb slowly for about 20 minutes before streaking and care should be taken to remove all the contents of the ampoule as the spores are usually at its base due to the centrifugal action.

(viii) Storage may be at room temperature (Plate 7) but lower storage temperatures such as 15°C, 4–8°C or even sub zero temperatures are preferred.

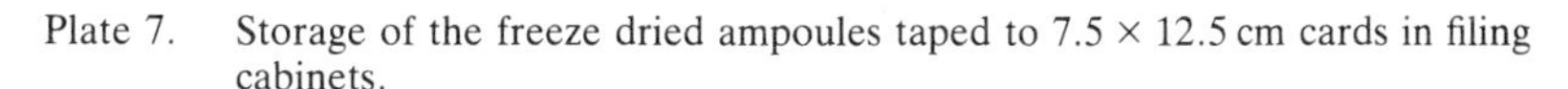

Plate 7. Storage of the freeze dried ampoules taped to 7.5 × 12.5 cm cards in filing cabinets.

Only sporulating fungi seem to survive centrifugal freeze drying though some sterile ascocarps, sclerotia and other resting stages have been processed successfully. The spores with high moisture content tend not to freeze dry by the centrifugal freeze drying technique though pre-freezing before vacuum drying has proved successful with some normally sensitive fungi (Smith, 1983a).

Results of the process are given as a taxonomic list of viability (Table 8), a list of some organisms that have survived without suspending media (Table 9) and survival periods (Table 10). A list of genera that have failed centrifugal freeze drying is given (Table 11).

Other methods of processing include the use of a shelf drier (Plate 8), in which the cooling can be controlled on a cold shelf inside the vacuum chamber. The vials used

Table 8 Viability of isolates after centrifugal freeze drying.

	Number of genera viable	Number species viable	Number of isolates		
			Tested	Viable	% Viability
Mastigomycotina					
Chytridiomycetes	0	0	6	0	0
Oomycetes	0	0	29	0	0
Zygomycotina					
Zygomycetes	48	222	821	754	92
Ascomycotina	203	692	1463	1228	84
Basidiomycotina	28	47	121	62	51
Deuteromycotina					
Hyphomycetes	320	1565	5091	4822	95
Coelomycetes	106	302	576	526	91

Summary of viabilities of the isolates tested:

Number tested	Number successful	% Viability
8107	7392	91

The taxonomic breakdown was made according to Hawksworth et al. (1983).

have grooved bungs to allow evacuation and drying and can be sealed mechanically by pressing home while still under vacuum. The latter overcomes the time-consuming constriction of the ampoules between the two stages which is necessary with the centrifugal driers. A simple form of freeze drying is used at the ATCC in which a batch of ampoules are prefrozen and the vessel containing them is attached to the vacuum and drying trap system (Alexander et al, 1980). On completion of the process the vessel is back filled with nitrogen and the ampoules sealed later. The presence of the nitrogen overcomes the danger of exposure to atmospheric oxygen.

Table 9 Survival periods of isolates centrifugally freeze dried on agar discs without suspending medium.

Isolate	IMI No.	Maximum survival period (years)
Acremonium sp.	55286	14
Ascocoryne sarcoides	68130	14
Aspergillus amstelodami	71295	8*
A. candidus	73074	14
A. carneus	73777	14
A. nidulans var *echinulatus*	61454ii	14
A. niger	75353ii	14
A. quadrilineatus	72733	14
A. ostianus	93445	14
Chaetomium abuense	114513	14
Curvularia trifolii f.sp. *gladioli*	75377	13
Cylindrocarpon congoense	69504	14
Embellisia chlamydospora	67737	14
Fusarium graminearum	69695	14
Nectria gliocladioides	71095	14
Paecilomyces dactylethromorphus	65752	14
Penicillium cyclopium var. *echinulatum*	68236	14
P. nigricans	96660	14
P. paraherquei	68220	14
P. raperi	71625	13
P. roquefortii	129207	14
P. spinuloramigenum	68617	14
P. steckii	72029	14
Pestalotiopsis gracilis	69749	14
Phaeotrichoconis crotalariae	69755	14
Phialomyces macrosporus	110130	14
Phomopsis oncostoma	68344	14
Pycnoporus sanguineus	75002	9
Sagenomella griseoviridis	113160	14
Scopulariopsis carbonaria	86941	14
Sporidesmium flexum	246524	2

* Died after 8 years storage all other isolates are still viable.

Table 10 A list of genera that have survived 13–15 years storage freeze dried.

Absidia	*Didymella*	*Paecilomyces*
Acremonium	*Didymosphaeria*	*Penicillium*
Acrophialophora	*Doratomyces*	*Pestalotiopsis*
Actinomucor	*Drechslera*	*Pseudallescheria*
Alternaria	*Eladia*	*Phaeotrichoconis*
Amorphotheca	*Elsinoë*	*Phialomyces*
Arthroderma	*Emericellopsis*	*Phialophora*
Ascotricha	*Eremothecium*	*Phoma*
Aspergillus	*Exophiala*	*Phomopsis*
Aureobasidium	*Fusarium*	*Phycomyces*
Beauveria	*Gelasinospora*	*Pirella*
Bispora	*Geomyces*	*Pithomyces*
Botryotrichum	*Geosmithia*	*Pleurographium*
Byssochlamys	*Geotrichum*	*Rhinotrichum*
Calcarisporium	*Gilbertella*	*Rhizopus*
Cephaliophora	*Gliocladium*	*Scolecobasidium*
Cercospora	*Gliomastix*	*Scopulariopsis*
Chaetodiplodia	*Glomerella*	*Scytalidium*
Chaetomella	*Gonytrichum*	*Sesquicillium*
Chaetomium	*Heimiodora*	*Setosphaeria*
Chaetospina	*Hirsutella*	*Spegazzinia*
Chalara	*Humicola*	*Sporothrix*
Chloridium	*Leptographium*	*Stachybotrys*
Chrysosporium	*Leptosphaerulina*	*Staphylotrichum*
Circinella	*Loramyces*	*Stemphylium*
Cladosporium	*Mammaria*	*Sydowia*
Claviceps	*Mariannaea*	*Syncephalastrum*
Cochliobolus	*Memnoniella*	*Thamnostylum*
Colletotrichum	*Microascus*	*Thermoascus*
Coniella	*Mucor*	*Thielavia*
Coniothyrium	*Myrothecium*	*Thysanophora*
Cordana	*Nectria*	*Trichoderma*
Cordyceps	*Neocosmospora*	*Ulocladium*
Coryne	*Neodeightonia*	*Venturia*
Cunninghamella	*Neurospora*	*Verticillium*
Curvularia	*Oidiodendron*	*Volutella*
Cylindrocarpon	*Ophiobolus*	*Zygorhynchus*

Advantages of freeze drying

(i) Total sealing of the specimen and protection from infection and infestation.
(ii) Very long viabilities.
(iii) Many isolates retained in a very stable condition.

Table 11 Genera that failed centrifugal freeze drying.

Achlya	*Kretzschmaria*	*Quarternaria*
Allomyces	*Lacellinopsis*	*Saprolegnia*
Areolospora	*Lasiobolidium*	*Searchomyces*
Armillaria	*Lentinus*	*Selenosporella*
Arthrocladium	*Lenzites*	*Selinia*
Ascocalvatia	*Leptoporus*	*Sigmoidea*
Balansia	*Lomachashaka*	*Sphaerobolus*
Batterraea	*Marasmius*	*Sphaerostilbe*
Blastocladiella	*Melanconis*	*Spondylocladiopsis*
Calospora	*Monotosporella*	*Stereum*
Camposporium	*Nummularia*	*Sympodiella*
Chytridium	*Panus*	*Syzygites*
Cladobotryum	*Penicillifer*	*Tetracladium*
Coriolus	*Phaeoisariopsis*	*Tetracrium*
Dactuliophora	*Phyllosticta*	*Umbelopsis*
Eleutherascus	*Physarum*	*Urohendersonia*
Entomophthora	*Phytophthora*	*Ustilaginoidea*
Eremomyces	*Piedraia*	*Ustulina*
Fomes	*Platystomum*	*Volvariella*
Ganoderma	*Puccinia*	
Herpotrichia	*Pythium*	

Disadvantages of freeze drying

(i) Some isolates fail to survive the process (Table 11).

(ii) Some methods produce a low percentage viability, therefore it is not usually suitable for starter cultures.

(iii) Genetic damage may occur (Ashwood-Smith & Grant, 1976) though unless high viability is retained it is difficult to differentiate between this and selection of spontaneous mutants by freeze drying (Heckley, 1978). With careful control of freezing protocol and prevention of overdrying during the process this can be avoided.

(iv) The process is complex and may be expensive and at least requires a vacuum system.

This technique has proved to be a most convenient and successful method of preserving sporulating fungi. It removes the possibility of contamination during storage and is also ideal for the distribution of the organism. The organisms are kept very stable in storage.

7. Liquid nitrogen storage

The storage of microorganisms at ultra low temperatures (−196°C for liquid nitrogen) is at present regarded as the ultimate method of preservation. It is a form of induced dormancy during which the organism does not undergo any change either phenotypically or genotypically providing adequate care is taken during freezing and thawing. The method can be applied to both sporulating and non-sporulating cultures and it is believed that the failure of a few groups and species to respond is only a matter of technique which can be eventually overcome.

At the CMI there are over 3,000 isolates in the liquid nitrogen collection, these comprise 2,075 species and 595 genera. No morphological or physiological change has been observed. The method involves preparation of a suspension of fungal spores or mycelium in a protective medium and freezing at 1°C/min at −35°C followed by rapid uncontrolled cooling to −196°C. Thawing is rapid.

Variations of the method are numerous. Different rates of cooling and thawing and the use of different cryoprotectants have all been tried. Initial work with fungi was

Plate 8. A shelf freeze drier which enables prefreezing before evacuation and drying and also allows sealing under vacuum in rubber stoppered vials without exposing the dried suspension to the atmosphere.

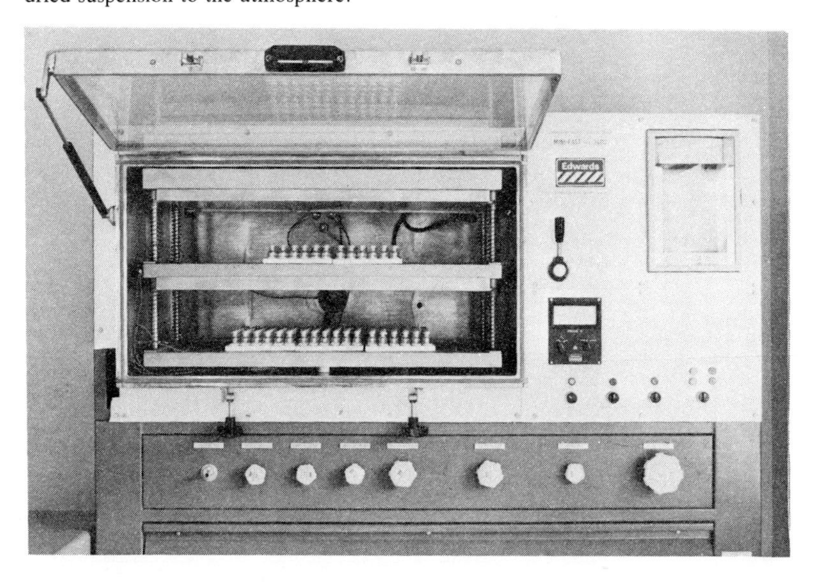

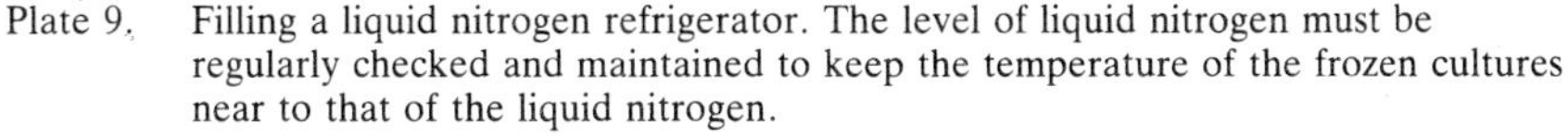

Plate 9. Filling a liquid nitrogen refrigerator. The level of liquid nitrogen must be regularly checked and maintained to keep the temperature of the frozen cultures near to that of the liquid nitrogen.

undertaken by Hwang (1960) at the American Type Culture Collection; further reports of methods used have been given (Calcott, 1978; Onions, 1971, 1983; Smith, 1982).

Factors and variation of method to be considered include:

(i) *Choice and size of refrigerators.* This depends on funds as these containers tend to be expensive (Plate 9).

(ii) *Choice of ampoules.* Glass ampoules are not always easy to handle; they may crack and it is difficult to avoid flaws in the seals. Faulty ampoules allow liquid nitrogen to enter and explode on removal from the refrigerator. Various plastic ampoules have been developed; these are usually best stored in the vapour phase (Butterfield et al, 1978; Simione et al, 1977). Other alternatives include special plastic drinking straws (Dietz, 1975; Elliott, 1976) (Plate 10) and plastic film (Tuite, 1968).

(iii) *Cultures and preparation of suspension.* Cultures should be healthy and exhibit all characters to be preserved, both morphologically and physiologically. Sterile cultures survive the technique well but it is desirable to allow full development before preservation. Sporulating cultures give better viabilities. Poor isolates will not be improved by the technique and may be more sensitive to the process. Many of the failures at CMI were due to cultures initially being in poor condition. Care must be taken not to cause mechanical damage during preparation of suspensions and various precautions to avoid this can be taken. Fungi on slivers of agar can be put into the ampoules or the fungus can be grown on small amounts of agar in the ampoule before the cryoprotectant is added.

(iv) *Cryoprotectant.* The choice of cryoprotectant is a matter of experience and varies according to the organism to be preserved. Glycerol often gives very satisfactory results, but needs time to penetrate the organisms; some organisms are damaged by this delay. Dimethyl sulphoxide (DMSO) penetrates rapidly and is often more satisfactory (Hwang & Howells, 1968; Hwang et al, 1976). Other protectants have been used including sugars and large molecular substances such as polyvinylpyrrolidone (PVP) (Ashwood-Smith & Warby, 1971). Mixtures of protectants have proved particularly satisfactory at CMI (Smith, 1983b).

Plate 10. A selection of ampoules and straws used at CMI for the preservation of fungi in liquid nitrogen. The 1 ml borosilicate glass ampoules are shown at the top of the photograph and some clipped to an aluminium cane below them. Plastic ampoules (bottom left) and plastic straws (bottom right) for storage in the vapour phase of the refrigerator.

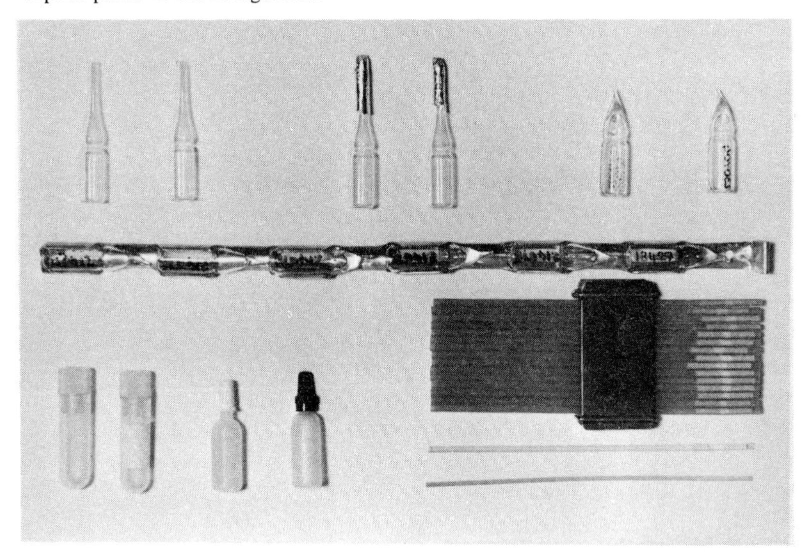

(v) *Cooling rate*. This has been the subject of much research. For fungi slow cooling at 1°C/min over the critical phase seems satisfactory (Hwang, 1966, 1968) but some isolates are not very sensitive and some respond well to ultra rapid cooling, preferably without protectant. Controlled cooling equipment can be complex and expensive, but a good approximation can often be worked out by suspending ampoules in the neck of the refrigerator.

(vi) *Thawing rate*. Slow thawing may cause damage due to recrystallization of ice during warming, therefore rapid thawing is recommended. Slow freezing and rapid thawing generally give the highest viabilities (Heckly, 1978).

(vii) *Storage*. Generally storage is at −196°C in the liquid nitrogen or slightly higher in the vapour phase. The development of evaporating freezers which can produce temperatures of −80°C or lower may result in their substitution in the future. At present, however, these are expensive and it is not proven whether such temperatures will maintain stability during storage.

(viii) *Filing*. The original ampoules were clipped on to aluminium canes, and stored vertically in metal boxes. More recently freeze file systems (Denley, 14) have evolved which make retrieval easier (Plate 11).

Method used at CMI

(i) Cold hardening of the cultures prior to freezing may prove beneficial. Pregrowth of cultures in the refrigerator (4–7°C) can give improved viabilities with some fungi though others cannot grow at low temperature. For those isolates that are sensitive a short exposure may suffice or this step may be omitted altogether.

(ii) A spore or mycelial suspension is prepared in 10% glycerol; mechanical damage must be avoided.

(iii) 0.5 ml aliquots are distributed in sterile 1 ml borosilicate glass ampoules (Anchor

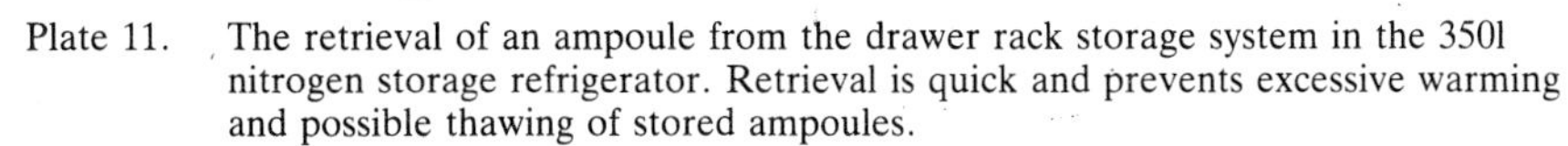
Plate 11. The retrieval of an ampoule from the drawer rack storage system in the 350l nitrogen storage refrigerator. Retrieval is quick and prevents excessive warming and possible thawing of stored ampoules.

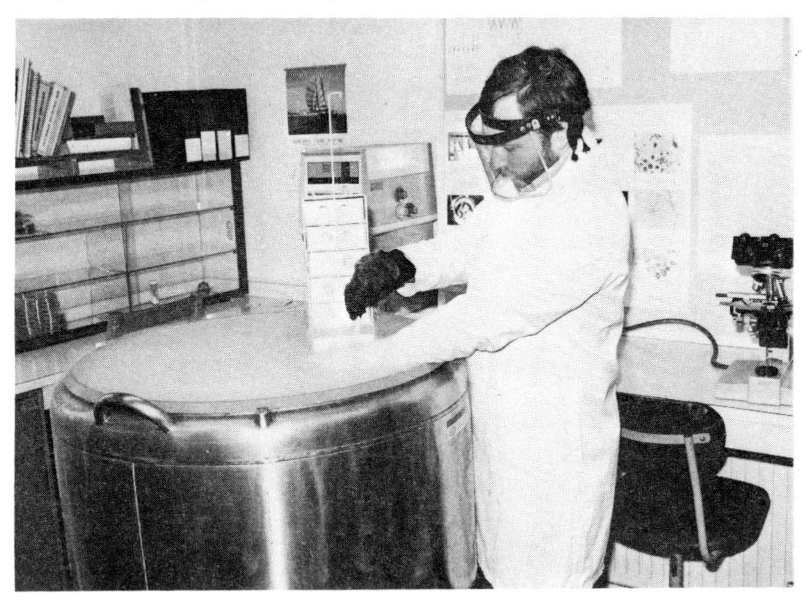

Glass Co. Ltd., 6), labelled with the culture number using a permanent ink marker (Scott's Office Equipment, 13).

(iv) The ampoules are sealed using an air/gas Flame master torch (Buck & Hickman, 3) and placed in a dye bath in a refrigerator 4–8°C for ½ hour to pre-cool. This may allow the glycerol to penetrate the organism and the dye to indicate any faulty seals in the ampoules by colouring the contents.

(v) The ampoules are then placed on canes and cooled at approximately 1°C/min in the neck of the nitrogen refrigerator (Union Carbide, 8) at −35°C. They are placed in boxes and lowered into liquid nitrogen and cooled to −196°C and stored in the main part of the refrigerator.

Table 12 Viability of isolates stored in liquid nitrogen for 2–13 years.

	Number of genera viable	Number of species viable	Number of isolates		
			Tested	Viable	% Viability
Mastigomycotina					
Chytridiomycetes	4	7	56	9	16
Hyphochytriomycetes	1	1	5	3	60
Oomycetes	9	54	348	172	50
Zygomycotina					
Zygomycetes	43	176	267	254	95
Ascomycotina	168	439	626	600	96
Basidiomycotina	48	75	190	167	88
Deuteromycotina					
Hyphomycetes	259	1212	1543	1465	95
Coelomycetes	67	163	238	224	94

Summary of the viabilities of the isolates tested:

Number tested	Number viable	% Viability
3273	2894	88

The taxonomic breakdown was made according to Hawksworth et al. (1983).

(vi) After 3 to 4 days storage an ampoule is thawed rapidly in a water bath at 37°C and the contents streaked out on a suitable medium to check viability and purity.

This process is very simple and gives good routine results (Table 12).

Advantages of liquid nitrogen storage

(i) Cultures are maintained in stable condition (Prescott & Kernkamp, 1971) over very long periods. Limits of viability have not yet been recorded.

(ii) The cultures can be completely sealed free from contamination.

(iii) The majority of both sporulating and non sporulating fungi survive well (Smith, 1982).

Table 13 A list of genera that have survived over 12 years storage in liquid nitrogen.

Absidia	*Farrowia*	*Piptocephalis*
Achaetomium	*Fusarium*	*Pithomyces*
Achlya	*Geniculosporium*	*Pleospora*
Acremonium	*Geosmithia*	*Podospora*
Acrostalagmus	*Geotrichum*	*Polyporus*
Alternaria	*Gliocephalotrichum*	*Pseudeurotium*
Alysidium	*Gliocladium*	*Pseudocersporella*
Anixiella	*Gliomastix*	*Pyrenophora*
Apiocrea	*Glomerella*	*Pythium*
Apiosordaria	*Gongronella*	*Rhinocladiella*
Armillaria	*Hansenula*	*Rhizomucor*
Arthrobotrys	*Heimiodora*	*Rhizopus*
Ascophanus	*Hesseltinella*	*Rhynchosporium*
Aspergillus	*Heterocephalum*	*Robillarda*
Backusella	*Humicola*	*Rosellinia*
Botryosporium	*Hyphochytrium*	*Ryparobius*
Botrytis	*Hypomyces*	*Saccharomyces*
Camarosporium	*Isaria*	*Schizosaccharomyces*
Ceratocystis	*Kabatiella*	*Schizotrichum*
Cercospora	*Keratinophyton*	*Sclerotinia*
Chaetomella	*Khuskia*	*Scolecobasidium*
Chaetomium	*Leptosphaeria*	*Scopulariopsis*
Chalaropsis	*Melanospora*	*Septoria*
Chlamydoabsidia	*Metarhizium*	*Sirodesmium*
Circinella	*Micropolyspora*	*Sordaria*
Cladosporium	*Mortierella*	*Sporotrichum*
Cleistothelebolus	*Mucor*	*Stachybotrys*
Cochliobolus	*Mycotypha*	*Stagonospora*
Coemansia	*Myrothecium*	*Stereum*
Cokeromyces	*Myxotrichum*	*Stigmina*
Colletotrichum	*Narasimhania*	*Streptotheca*
Conidiobolus	*Nectria*	*Stromatinia*
Coniella	*Neurospora*	*Taphrina*
Coprinus	*Nigrosabulum*	*Thamnidium*
Corticium	*Nodulisporium*	*Thanatephorus*
Cunninghamella	*Oedocephalum*	*Thermoascus*
Curvularia	*Paecilomyces*	*Thermomyces*
Dactuliophora	*Penicillium*	*Thielavia*
Dactylella	*Periconia*	*Tilletiopsis*
Dendrosporium	*Periconiella*	*Torula*
Diaporthe	*Pestalotiopsis*	*Torulomyces*
Dichotomomyces	*Peziza*	*Trichoderma*
Dictyoarthrinium	*Phaeopolynema*	*Trichurus*
Dictyostelium	*Phialophora*	*Valsa*
Drechslera	*Phialotubus*	*Volutella*
Entyloma	*Phlyctochytrium*	*Zopfiella*
Eremodothis	*Phoma*	*Zygorhynchus*
Eupenicillium	*Phytophthora*	
Exophiala	*Pilaira*	

Disadvantages of liquid nitrogen storage

(i) The apparatus is expensive.

(ii) A continuous supply of nitrogen is expensive.

(iii) A continuous supply of nitrogen is required. If the supply fails then the collection is lost. The ATCC have an elaborate warning system to detect refrigerator failure or loss of nitrogen. A secondary support collection is almost a necessity.

Liquid nitrogen is probably the technique that can be most widely applied to the storage of fungi. The viabilities of Zygomycetes, Ascomycetes, Basidiomycetes, Hyphomycetes and Coelomycetes after freezing and storage are very high (Table 12). The Mastigomycotina did less well though those that failed did not grow well in culture and therefore would not be expected to improve and give good results after storage. No tables showing failures were included as experience has shown that the results were due to the poor isolates tested. One hundred and forty five genera are listed (Table 13) that have survived 12 years or more though it is expected that these fungi will survive much longer periods than have been observed under oil at CMI (32 years).

This method should be the technique of choice for all fungi if funds and facilities are available. The potential storage periods and the stability of the fungi stored are unequalled by any of the other techniques described here.

III Comparison of Methods of Preservation

The choice of method of preservation depends upon the requirements of the collection. These vary according to the numbers and range of fungi to be preserved and the facilities available. The cost of materials and labour involved and the desired level of stability and longevity required must be taken into consideration.

Collection Requirements

Culture collections vary in size and function. They may be private collections belonging to a university department, or a school, holding small numbers of a wide variety of common organisms maintained for teaching purposes. Such a collection is useful rather than valuable and the isolates can be easily replaced from other collections or reisolated from the original habitat. Little time, money or labour can be spared so methods employed should be simple and cheap. Longevity is not a major concern and stability though desirable is not essential. Oil storage and the deep freeze should be considered here.

Other private collections built up in support of research projects are often very specific. They may consist of many isolates of one species or only a few isolates, which are of great importance to the researcher and may be required in genetic or biochemical studies. This material must be kept unchanged and is of potential long term interest to the community. The actual project may be relatively short, but at its completion representative or important isolates should be deposited with one of the national collections. Individual isolates of great value should perhaps be deposited during the work to protect the researcher from accidental loss. Such specialist collections may not require long term preservation but stability is very important. Time and technical assistance may not be available, funds may be limited, and labour may be at a premium, so methods such as deep freeze or silica gel storage might prove useful. Regular transfer should still be considered when dealing with a few specialists isolates and careful curation by researchers has kept many valuable private collections in good condition by this means. Though freeze drying and liquid nitrogen are preferable techniques for such a collection they may not be practical because of the expense involved. However, the facilities may be available, for example in another university department, and could prove to be a cheap and reliable means of preservation. Some collections are set up to support a service or an industrial project. They may be required to supply living starter material at regular intervals, in industrial fermentation for example. The culture must always be readily available, active and absolutely stable. The cost in money, equipment and labour is probably small in relation to the importance of stability and availability. To avoid accidental loss several methods of maintenance may be used including deposit in a national collection. Regular transfer on suitable media may suffice for the day to day use of the organism or for the build up of inoculum. The techniques of freeze drying, liquid nitrogen or silica gel storage will serve as a stable back up of stock cultures for replacement of organisms that may suffer variation, contamination or that may die during transfer and use.

Finally the service collections facilitate the supply of cultures to the public. These are

Table 14 Culture collection requirements

Type of Collection	Requirements						Methods
	Access required	Numbers of isolates	Stability	Longevity	Cost	Labour available	
Private (University)							
Teaching	easy	low	low	low	low	low	Oil, Water, Soil, SG, Deep freeze
Research (few important isolates)	easy	low	high	medium	medium	relatively high	FD & LN if available, SG, Deep freeze
Research (many similar isolates)	easy	high	medium/ high	medium	low	low	Deep freeze, Oil, SG
Working							
Industrial	quick and easy	high/low	very high	medium/ high	high	high	FD, LN, SG, Deep freeze
Institutional	easy	possibly high	varies	medium	low	low	Depends on availability, FD, LN, SG, Deep freeze
Service (National)	easy	high	high	high	high	low (but high for individuals)	All techniques

SG, silica gel storage FD, freeze drying LN, liquid nitrogen storage

often supported on a national basis and give a variety of additional services such as identification of organisms, and they provide a depository for types or patent application cultures. Many organisms of differing requirements have to be maintained in a highly stable condition for long periods. Specialist supervision is essential and several methods of preservation are desirable, including at least one of high stability and considerable longevity such as liquid nitrogen or freeze drying, with a back up of other methods as necessary. There will be some highly sensitive isolates that will require specialist treatment. The relative cost per isolate, despite the sophisticated methods, may not be excessive due to the scale of the operation. Table 14 compares types of collections and their needs.

Comparison of Methods Available

When choosing a method of preservation there must be a balance between requirements, cost in material and labour, longevity and genetic stability. Table 15 compares these factors for different methods of preservation. However, these are not the only factors involved. Table 16 compares the effectiveness of different methods of preservation for different organisms. This is given in general terms and individual species and isolates do not necessarily conform to the general rule.

In most instances liquid nitrogen storage gives the most reliable results. However, for various reasons this method may not be suitable and it is so sensitive to failure of the supply of nitrogen that in almost every case there should be a back up of some other preservation method. Though the stability and longevity of freeze drying is perhaps not as great, this method is not vulnerable in the way that liquid nitrogen storage is, but it is initially technically complex and expensive. Not all organisms freeze dry and some workers cast doubt on its giving reliable genetic stability. However, it must be emphasized that preservation techniques seldom improve an isolate and many

Table 15 A comparison of methods of preservation.

Method of Preservation	Cost		Longevity	Genetic Stability	General Comment
	Material	Labour			
Regular transfer on agar					
(i) Storage at room temperature	low	high	1–6 months	variable, depends on curation	Keep stock culture in case of contamination of working culture
(ii) Storage in the refrigerator	*medium	high	6–12 months	variable	
(iii) Storage under oil	low	low/medium	1–32 years	poor	
(iv) Storage in water	low	low/medium	2–5 years	moderate	
(v) Storage in the deep freeze	*medium	low/medium	4–5 years	moderate	Must not allow thawing of material while subbing
Drying					
In soil	low	medium	5–20 years	moderate to low	
Silica gel	low	medium	5–11 years	good	
Freeze drying	high	† initially medium	4–40 years	good	Recent work reports that overdrying can cause DNA damage
Freezing					
Liquid nitrogen storage	high	low	infinite: 14 years to date at CMI	good	

* Refrigerator and deep freeze costs included.
† Initial processing costly depending on the method though maintenance is negligible.

reported failures especially for freeze drying and even for liquid nitrogen result from processing already altered material.

The early collections were maintained for years by frequent transfer and specialist curation. Several cultures in the CMI collection were once in the collections of Krall and Thom and date back to between 1900 and 1910. To reduce the labour and time involved, covering the growing cultures with a layer of mineral oil became a well used method. Mineral oil cultures are generally unpopular as being messy and encouraging mutations but the method is cheap, it deters mites, and some fungi at the CMI have survived 32 years storage, a not insignificant record; so oil storage is a method which should not be overlooked. Water storage appears promising, fairly stable and more pleasant to handle, though survival is for a shorter period. Although some fungi will

Table 16 Preservation techniques for a range of fungi

	Transfer on agar					Soil	Silica gel	Freeze drying	Liquid nitrogen
	Room temperature	Refrigerator	Mineral Oil	Water	Deep freeze				
Mastigomycotina (excluding Oomycetes)	fair	**fair	fair	poor	poor	fair	fail	fail	fair
Oomycetes	poor	poor	fair/good	good	poor	fair	fail	†poor	good
Zygomycotina (excluding Entomophthorales)	good	**fair	fair	good	fair	good	good	v. good	v. good
Entomophthorales	poor	poor	good	fair	poor	fair	fail	fail	good
Ascomycotina (excluding Laboulbeniales)	good	good	fair	good	good	good	v. good	v. good	v. good
*Laboulbeniales	poor	poor	poor	poor	poor	poor	fail	poor	good
Basidiomycotina									
(a) mycelial in culture	fair	good	good	good	good	good	fail	fail	good
(b) Those that sporulate in culture	fair	fair	good	good	good	good	good	good	v. good
*(Rusts) Uredinales	fail	fail	fail	fail	poor	poor	fail	poor	good
*(Smuts) Ustilaginales	fair	fair	fair	fail	poor	poor	fail	fair	good
Deuteromycotina	fair	good	poor/good (variation)	good	good	good	v. good	v. good	v. good

* Experience at CMI is limited
** Some fail
† Some species of *Pythium* and *Phytophthora* survive

survive longer in the deep freeze than they do in water this method is still short term. However it is very convenient. Storage in silica gel or soil is useful, though the stability of fungi stored in silica gel has not been fully investigated. Experience with the technique at CMI shows that the isolates tested remained stable for periods of up to 11 years. Longevities for freeze drying and liquid nitrogen storage can be misleading as their limits of viability have not yet been recorded.

Suitability of the Techniques for Fungi

No one technique has been successfully applied to all fungi. There seem to be sensitive strains associated with every preservation method though storage in liquid nitrogen appears to approach the ideal. All fungi that grow well in culture survive freezing in liquid nitrogen. Those that grow poorly tend to do less well, even non-sporulating fungi can survive. This technique will also allow the preservation of fungal pathogens in infected tissue bearing in mind that a mixed population of organisms may be preserved. It is a method that enables the preservation of fungi that do not grow in culture.

Of the remaining techniques described here, storage in mineral oil, soil or water is of use for a wide range of organisms. It seems that the majority survive but the period of storage for some may be very short. Only sporulating fungi survive storage in silica gel, and spores with thin walls and high water content or those with appendages do less well. Centrifugal freeze drying allows only the more robust spores to survive. Some sclerotia and other resting stages, and even in a few cases sterile mycelia, have been known to survive freeze drying. Mastigomycotina and oomycetes are very vulnerable to harsh treatment, but will survive long periods in mineral oil, in water or by frequent transfer. Most will not, however, survive drying in silica gel or freeze drying. Entomophthorales do well in liquid nitrogen but will also survive in oil or in soil. Non-sporulating Basidiomycetes can be successfully stored in liquid nitrogen. However, several large collections of such fungi have been successfully maintained by frequent transfer, under mineral oil or by refrigeration and it seems likely they might do well in water or in soil.

Small collections of teaching cultures may have a wide range of organisms. If the numbers involved are very small, frequent transfer may not prove too time consuming. Putting the cultures in the deep freeze may prove effective while slightly more permanent stocks could be kept in silica gel, especially those to be used to demonstrate genetic activity.

Rusts do not normally grow in culture but living collections can be maintained in good condition in liquid nitrogen. Ustilaginales produce very disappointing cultures and survive best in liquid nitrogen though it is possible to keep them by other means. If the spores can be harvested successfully some survive freeze drying quite well.

There are fungi that change or deteriorate quite quickly when grown in culture and regularly subcultured, such as species of *Fusarium*. Such fungi should be kept in a manner that will avoid serial transfers, and silica gel, water and soil storage are simple techniques to use if freeze drying and liquid nitrogen are not suitable.

Appendix 1

Formulae for media

Cornmeal (Maize) Agar (CMA)

Maize .... 30 g
Agar .... 20 g
Water .... 1 l
Place the maize and water in the saucepan (if meal is not available break up 30–35 g of grain and pass through the coffee mill). Heat over the double saucepan until boiling, stirring for 1 hour. Filter the decoction through muslin, add the agar, and heat until it is dissolved. Autoclave for 15 minutes at 121°C.

Czapek (Dox) Agar (CZ)

Made with Stock Czapek Solutions
Solutions A and B

A
Sodium nitrate $NaNO_3$ .... 40 g
Potassium chloride KCl .... 10 g
Magnesium sulphate $MgSO_4$ $7H_2O$ .... 10 g
Ferrous sulphate $FeSO_4$ $7H_2O$ .... 0.2 g
Dissolve in 1 l distilled water and store in a refrigerator.

B
Di-potassium hydrogen ortho phosphate K_2HPO_4 .. 20 g
Dissolve in 1 l distilled water and store in a refrigerator.

For 1 litre
Stock Solution A .... 50 ml
Stock Solution B .... 50 ml
Distilled water .... 900 ml
Sucrose .... 30 g
Agar .... 20 g
To a litre add 1.0 ml of both a and b:—
(a) Zinc sulphate $ZnSO_4$ $7H_2O$ .... 1.0 g in 100 ml water
(b) Copper sulphate $CuSO_4$ $5H_2O$ .... 0.5 g in 100 ml water
Autoclave for 20 minutes at 121°C.

Czapek Yeast Autolysate (CYA)

Czapek concentrate

Sodium nitrate $NaNO_3$	30 g
Potassium chloride KCl	5 g
Magnesium sulphate $MgSO_4 \cdot 7H_2O$	5 g
Ferrous sulphate $FeSO_4 \cdot 7H_2O$	0.1 g
Water	100 ml

Di-potassium hydrogen ortho phosphate K_2HPO_4	1.0 g
Czapek concentrate (see earlier)	10 ml
Yeast extract or autolysate	5.0 g
Sucrose	30 g
Agar	15 g
Distilled water	1 l

Autoclave for 15 minutes at 121°C.
For *Penicillium* identifications (Pitt, 1973, 1979).

Egg Yolk Medium

Soak eggs in 90% alcohol, with a little acetone added, for 2 hours.
Flame off, puncture a (5 mm) hole into each end, pour white away.
Puncture yolk membrane, pour into bottles or plates.
Steam at 80°C for 30 to 45 minutes.
Entomophthora and *Conidiobolus* species will survive a few transfers on Sabouraud agar.

Malt Czapek Agar (MCz)

Use solutions A and B as described earlier

Stock Czapek solution A	50 ml
Stock Czapek solution B	50 ml
Sucrose	30 g
Malt extract	40 g
Agar	20 g
Distilled water	900 ml

Dissolve malt extract and agar in water. Add the two stock solutions and sucrose and heat over double saucepan until dissolved. Dispense into bottles and autoclave for 20 minutes at 121°C. Adjust pH to between 4 and 5.

Malt Extract Agar (MEA)

Malt extract (powdered, Difco or Oxoid)	20 g
Peptone (bacteriological)	1.0 g
Glucose	20 g
Agar	15 g
Distilled water	1 l

Autoclave for 15 minutes at 121°C.
For *Penicillium* identifications (Raper & Thom, 1949; Pitt, 1973)

Malt Extract with added sucrose (for organisms requiring high osmotic pressure for sporulation).

	M/20	M/40	M/60
Malt extract	20 g	20 g	20 g
Sucrose	200 g	400 g	600 g
agar	20 g	20 g	20 g
Water	1 l	1 l	1 l

Prepare in the same way as Malt extract agar, but add sugar just before boiling to reduce caramelization.

Malt Extract Agar (MA)

Malt extract (good quality) .... 20 g
Agar .... 20 g
Water .... 1 l

Boil the malt extract in water until dissolved; add agar and boil until agar is all dissolved. Dispense as required, autoclave for 20 minutes at 121°C.

The brand of malt extract appears to be an important factor with this medium, for success that supplied by Edme Ltd. (16) is satisfactory. Do not filter after adding the malt. The pH will be between 3 and 4, this should be adjusted to 6.5 with sodium hydroxide solution.

Oat Agar (OA)

Weigh out 30 g of powdered oatmeal, and add to 1 l of water in a saucepan. Gradually heat to boiling over a double saucepan, stirring occasionally. Simmer for 1 hour.

Pass through muslin, make up to 1 l, add 20 g of agar, and heat until dissolved. (Add 0.5 ml wheatgerm oil if required) and stir thoroughly. Dispense as required and autoclave for 20 minutes at 121°C. Use Japanese Kobe agar.

Potato Carrot Agar (PCA)

Wash, peel and grate potatoes and carrots as required.
For 1 litre
Weigh out 20 g of grated potato and 2 g of grated carrot. Boil vegetables for about 1 hour in 1 l of tap water, then drain through a fine sieve and add 20 g agar. Heat in a double saucepan until agar is dissolved. Pour into funnel and run into bottles. Sterilize in autoclave for 20 minutes at 121°C.

Potato Dextrose Agar (PDA)

Scrub, potatoes clean, and cut up without peeling into 12 mm cubes. Weigh out 200 g, and rinse rapidly under a running tap, and put in to 1 l of water in a saucepan. Boil until potatoes are soft (about 1 hour) then filter through a sieve and squeeze through as much pulp as possible. Alternatively blend, in which case a thicker opaque extract is produced. Add 20 g of agar, and heat until it dissolves. Add 15 g of dextrose and stir until it dissolves. Make up to 1 l.

Dispense into required containers agitating stock to ensure that the solid matter is evenly distributed. Sterilize in autoclave for 20 minutes at 121°C.
For PCA, PDA and PSA avoid new potatoes, which do not make good media.
Make large stocks before old potatoes disappear in early summer.

Potato Sucrose Agar (PSA)

Suspend 1.8 kg potatoes, peeled and diced, in double cheesecloth in 4½ l of water and boil until potatoes are almost cooked (about 8 minutes).
To make 1 litre of medium.
Potato water .... 500 ml
Sucrose .... 20 g
Agar .... 20 g
Distilled water .... 500 ml
Cook in double saucepan until agar is dissolved. Autoclave for 15 minutes at 121°C.
pH 6.5, adjusted with calcium carbonate if necessary.

Rabbit Dung Agar (RDA)

Place *5 pellets of the dung from wild rabbits in each bottle, and run in plain water agar.
Autoclave for 20 minutes at 121°C.
*6 pellets for plates.

Sabouraud's Agar

Sabouraud's Conservation Agar (SCA)
Peptone .... 10 g
Agar .... 20 g
Water .... 1 l
Autoclave for 20 minutes at 121°C.

Sabouraud's Test Agar (Glucose or Maltose)
*Peptone (Chassaing's) .... 30 g
Dextrose (or maltose) .... 40 g
Agar .... 20 g
Water .... 1 l
Autoclave for 20 minutes at 121°C.
Conservation medium delays pleomorphism.
* Sabouraud specified crude sugars containing essential polysaccharides. This is important. Sabouraud prescribed "glucose masse de Chanut" and "Maltose brut de Chanut". These crude sugars are no longer obtainable, and refined dextrose must be used; in this case it is essential to use Chassaing's peptone.

Soil Extract Agar (SEA)

(Flentje's formula for *Corticium praticola*, promoting formation of basidia).
Soil .... 1 kg
Water .... 1 l
This mixture should be agitated frequently for a day or two; pour extract through glass wool, and make up to 1 l.
Extract (prepared as above) .... 1 l
Sucrose .... 1 g
KH_2PO_4 .... 0.2 g

Dried yeast .... 0.1 g
Agar .... 25 g
Place all ingredients in double saucepan and boil until agar is dissolved. Dispense and sterilize by autoclaving for 20 minutes at 121°C.

Starch Agar (SA)

Soluble starch .... 40 g
Marmite .... 5 g
Agar .... 20 g
Tap water .... 1 l
Place all the constituents in to the water, and heat in the double saucepan, until dissolved. Dispense as required, and autoclave for 20 minutes at 121°C. pH is usually between 6.5–7 and requires no adjustment.

(Corn) Steep Agar

Corn steep liquor .... 90 ml
Agar .... 25 g
Water .... 1 l
Dissolve agar in the liquor and water and autoclave for 20 minutes at 121°C.

Tap water agar (TWA)

Tap water .... 1 l
Agar .... 15 g
Dissolve agar in water, dispense as required and autoclave for 20 minutes at 121°C.

V8 Juice Agar (V8)

V8 vegetable juice .... 200 ml (355 ml)
Agar .... 20 g (35 g)
Water .... 800 ml (1400 ml)
Dissolve agar in water and add vegetable juice. Adjust pH to 6.0 with 10% sodium hydroxide. Autoclave for 20 minutes at 121°C, pH after autoclaving should be 5.8.
Ref: Diener, 1955.

V8 Agar (as recommended for Actinomycetes).

V8 juice .... 200 ml (355 ml)
Calcium carbonate .... 4.0 g (7 g)
Agar .... 20 g (35 g)
Water .... 800 ml (1400 ml)
(Adjust to pH 7.3 with potassium hydroxide).
Ref: Galindo & Gallegly, 1960.
Amounts in brackets are given for a 355 ml can of V8 juice.

Yeast Peptone Soluble Starch (YPSS)

"Difco Yeast Extract"	4 g
Soluble starch	15 g
Di-potassium hydrogen ortho phosphate K_2HPO_4	1 g
Magnesium sulphate crystals $MgSO_4$ $7HO_2$	0.5 g
Agar	20 g
Water	1 l

Mix together, dissolve and dispense. Autoclave for 15 minutes at 121°C.

Appendix 2.

Growth conditions and preservation techniques of some common fungi.

	Growth media	Growth temp.	Storage techniques
Absidia glauca	PDA MA	RmT	R-FD, LN, SG. U-Oil, Soil, H_2O
A. spinosa	MA PDA	RmT	R-FD, LN, SG. U-Oil, Soil, H_2O
Achlya racemosa	OA, hemp seed	15°C	U-Oil, H_2O
Albugo candida	Host	suitable for host	R-LN, U-FD for spores
Allomyces arbuscula	OA YPSS	RmT	U-Oil, H_2O
A. javanicus	OA YPSS	RmT	U-Oil, H_2O
Alternaria alternata	MA PDA OA PCA	RmT BL	R-FD, SG, LN. U-Oil, Sol, H_2O
Armillaria mellea	MA PDA CMA	RmT	R-N. U-Oil
Arthrobotrys oligospora	RDA PCA	RmT	R-FD, LN. U-Oil
Ascobolus viridulus	RDA PCA + filter paper	RmT, direct sunlight	R-FD, SG, LN. U-Oil
Asochyta pisi	PCA PDA	RmT BL	R-FD, LN. U-SG, Oil
Aspergillus amstelodami	MA M_{20} PDA MCz Cz	RmT	R-FD, LN, SG. U-Oil
A. flavus	MA PDA MCz Cz	RmT	R-FD, LN, SG. U-Oil
A. fumigatus	MA PDA MCz Cz	RmT	R-FD, LN, SG. U-Oil
A. niger	MA PDA MCz Cz	RmT	R-FD, LN, SG. U-Oil
A. variecolor	MA PDA MCz Cz	RmT	R-FD, LN, SG. U-Oil
Aureobasidium pullulans	MA PDA	RmT	R-FD, LN. U-SG, Oil
Botryodiplodia theobromae	OA CMA	RmT BL	R-LN. U-FD, Oil
Botrytis cinerea	MA PDA OA	RmT	R-FD, LN, SG. U-Oil
Byssochlamys fulva	MA PDA	37°C	R-FD, LN, SG. U-Oil
Camarosporium aequivocum	PDA OA MA	RmT	R-FD, LN. U-Oil
Cephalotrichum stemonitis	MA PDA	RmT	R-FD, LN, SG. U-Oil
Ceratocystis fimbriata	CMA OA PCA + filter paper	RmT	R-FD, LN. U-Oil, SG
C. ulmi	CMA OA PCA + filter paper	RmT BL	R-FD, LN. U-Oil
Cercospora apii	V8 PDA Sporulates at low temperature on weak media	RmT BL	R-FD, LN. U-Oil (2 yr)
Chaetocladium brefeldii	PDA MA	15°C	R-FD, LN. U-Oil
Chaetomium globosum	MA PDA PCA + filter paper	RmT	R-FD, LN, SG. U-Oil, H_2O, Soil
Choanephora circinans	RDA MA PDA	RmT	R-FD, LN. U-SG, Oil
C. curcurbitarum	RDA MA PDA	RmT	R-FD, LN. U-SG, Oil
Cladosporium cladosporioides	PDA MA	RmT	R-FD, LN, SG. U-Oil, Soil, H_2O
C. herbarum	PDA MA	RmT	R-FD, LN, SG. U-Oil, Soil, H_2O
Claviceps purpurea	OA PDA	RmT	R-FD, LN. U-SG, Oil
Cochliobolus cymbopogonis	PCA TWA + W	RmT BL	R-FD, LN. U-Oil
Colletotrichum coccodes	PDA PCA	RmT BL	R-FD, LN
C. musae	PDA PCA	RmT BL	R-FD, LN
Conidiobolus coronata	MA PDA	RmT	R-LN. U-Oil
Coniochaeta hansenii	CMA PDA PCA + filter paper	RmT	R-FD, LN. U-Oil
Coniophora puteana	MA PDA	RmT	R-LN. U-Oil
Coprinus cinereus	RDA MA	RmT dark	R-LN, U-FD, Oil
Coriolus versicolor	MA PDA	RmT	R-LN. U-Oil
Cunninghamella elegans	PDA MA	RmT	R-FD, LN. U-SG Oil
Curvularia oryzae	PDA PCA TWA + W	RmT BL	R-FD, LN, SG. U-Oil
Dictyostelium discoideum	RDA CMA (bacteria)	RmT	R-FD, LN. U-Oil
Dipodascus albidus	MA YPSS	RmT	R-FD, LN. U-Oil
Drechslera heveae	PDA PCA TWA + W	RmT BL	R-FD, LN. U-Oil, SG
D. sorghicola	PDA PCA TWA + W	RmT BL	R-FD, LN. U-Oil, SG
Eremascus albus	M_{20} M_{40} MCz	RmT	R-FD, LN. U-Oil

	Growth media	Growth temp.	Storage techniques
Eremothecium ashbyi	MA PCA SA	RmT	R-FD, LN. U-Oil. Regular subbing
Erysiphe graminis	Host	Temp. suitable for host	R-LN
Fusarium solani	PDA MA PSA	RmT BL	R-FD, LN, SG. U-Soil
F. oxysporum	PDA MA PSA	RmT BL	R-FD, LN, SG. U-Soil
Geotrichum candidum	MA PDA	RmT	R-FD, LN, SG. U-Oil
Glomerella cingulata	PDA PCA YPSS	RmT	R-FD, LN. U-SG, Oil
Graphium putredinis	MA PDA CMA	RmT	R-FD, LN, SG. U-Oil
Guignardia musae	PDA PCA OA	RmT BL	R-FD, LN. U-Oil, SG
Linderina pennispora	YPSS MA	RmT	R-FD, LN. U-SG, Oil
Macrophomina phaseoli	PDA TWA + leaves	RmT BL	R-LN. U-FD, Oil
Moniliella acetoabutans	PDA OA	RmT	R-FD, LN, SG. U-Oil
Mortierella rammanniana	PDA, MA, SEA	RmT	R-FD, LN. U-Oil (2 yr)
M. polycephala	PDA MA SEA	RmT	R-FD, LN. U-Oil (2 yr)
Mucor circinelloides	PDA MA	RmT	R-FD, LN. U-Oil, SG
M. hiemalis	PDA MA	RmT	R-FD, LN. U-Oil, SG
M. racemosus	PDA MA	RmT	R-FD, LN. U-Oil, SG
Myxotrichum deflexum	MA PDA PCA	RmT BL	R-LN. U-FD, Oil (2 yr)
Nectria galligena	MA PCA PSA	RmT BL	R-FD, LN. U-Soil, Oil
N. inventa	MA PCA PSA	RmT BL	R-FD, LN. U-Soil, Oil
Neocosmospora vasinfecta	PDA PCA	RmT BL	R-LN. U-FD, Soil
Neurospora crassa	PDA PCA	RmT	R-FD, LN. U-SG
N. sitophila	PDA PCA	RmT	R-FD, LN
Paecilomyces variotii	PDA MA Cz Mz	RmT	R-FD, LN, SG. U-Oil
Penicillium chrysogenum	PDA MA Cz MCz	RmT	R-FD, LN, SG. U-Soil, H_2O, Oil
P. claviforme	PDA MA MCz Cz	RmT	R-FD, LN, SG. U-Oil, Soil, H_2O
P. cyclopium	PDA MA MCz Cz	RmT	R-FD, LN, SG. U-Oil, Soil, H_2O DF
P. expansum	PDA MA MCz Cz	RmT	R-FD, LN, SG. U-Oil, Soil, H_2O, DF
P. wortmanni	PDA MA MCz Cz	RmT	R-FD, LN, SG. U-Oil, Soil, H_2O, DF
Pestalotiopsis versicolor	PDA PCA MA	RmT BL	R-LN, FD. U-Oil
Phoma glomerata	MA PDA	RmT BL	R-FD, LN, SG. U-Oil
P. violacea	MA PDA	RmT BL	R-FD, LN, SG. U-Oil
Phomopsis conorum	MA PDA	RmT BL	R-FD, LN. U-SG, Oil
Phycomyces blakesleeanus	PDA MA	RmT	R-FD, LN, SG. U-Oil
Phytophthora cactorum	OA CMA	RmT	R-LN. U-Oil, H_2O
P. erythroseptica	OA CMA	RmT	R-LN. U-Oil, H_2O
P. infestans	OA CMA	RmT	U-LN, H_2O, Oil
P. megasperma	OA CMA	RmT	U-H_2O, Oil, LN
P. nicotianae	OA CMA	RmT	R-LN. U-Oil, H_2O
Pilaira anomala	PDA MA	RmT	R-FD, LN. U-Oil
Pilobolus sphaerosporus	RDA MA	RmT	R-LN, U-FD
Piptocephalis arrhiza	MA PDA	RmT	R-FD, LN. U-Oil, SG
P. freseniana	MA PDA	RmT	R-FD, LN. U-SG, Oil
Piptoporus betulinus	MA PDA	RmT	R-LN. U-Oil
Pleospora herbarum	MA PCA OA	RmT	R-FD, LN. U-SG, Oil
Podospora curvispora	RDA CMA	RmT BL	R-FD, LN. U-Oil
Pseudeurotium zonatum	MA OA	RmT BL	R-FD, LN. U-SG, Oil
Puccinia graminis	Host	Temp. suitable for host	R-LN Some uredospores have been FD
Pyronema domesticum	PCA PDA	RmT direct sunlight	R-LN, FD. U-Oil
Pythium debaryanum	OA CMA zoospores in dw	RmT	R-LN, U-Oil, H_2O
P. middletonii	OA CMA zoospores in dw	RmT	R-LN, U-Oil, H_2O
P. ultimum	OA CMA zoospores in dw	RmT	U-H_2O, Oil
Rhizomucor pusillus	PDA PCA	37°C	R-FD, LN. U-SG
Rizhophydium sphaerotheca	OA hemp seed	RmT	U-H_2O, Oil, LN
R. oryzae	PDA MA	RmT	R-FD, LN, SG. U-Soil
Rhizopus sexualis	MA PDA	RmT	R-FD, LN, SG. U-Soil
R. stolonifer	MA PDA	RmT	R-FD, LN, SG. U-Soil
Saprolegnia ferax	OA hemp seed	RmT	U-LN, H_2O, Oil
Schizophyllum commune	MA PDA	RmT	R-LN. U-Oil
Sclerospora graminis	Host	25°C	R-LN
Sclerotinia fructigena	PDA OA	RmT	R-FD, LN, SG. U-Oil
Septoria apiicola	PDA OA	RmT BL	R-FD, LN. U-Oil, SG
S. chrysanthemella	PDA OA	RmT BL	R-FD, LN. U-Oil, SG
Serpula lacrimans	MA PDA	Below 25°C	R-LN. U-Oil
Sodaria fimicola	MA RDA	RmT	R-FD, LN, SG. U-Oil
Sphaerobolus stellatus	OA CMA	RmT sunlight	R-LN. U-Oil
Sporobolomyces roseus	MA PDA	RmT	R-FD, LN, SG. U-Oil
Stachybotrys atra	PCA + filter paper	RmT	R-FD, LN, SG. U-Oil
Stereum purpureum	MA PDA	RmT	R-LN. U-Oil

	Growth media	Growth temp.	Storage techniques
Syncephalastrum racemosum	MA PDA	RmT	R-FD, LN. U-Oil
Syzygites megalocarpus	PDA RDA	RmT	R-LN. U-Oil (2 yr), FD, SG
Taphrina deformans	PDA MA	RmT	R-LN. U-FD, Oil
Thamnidium elegans	MA PDA	RmT	R-FD, LN, SG. U-Oil
Thanatephorus cucumeris	MA PDA RDA	RmT	R-LN. U-Oil
Thielavia albomyces	YPSS PCA OA MA	37–45°C	R-FD, LN, SG. U-Oil
Thraustotheca clavata	OA hemp or onion seed	RmT	U-H_2O, Oil
Trichoderma harzianum	MA PDA PCA + filter paper	RmT	R-FD, LN, SG. U-Oil
T. viride	MA PDA PCA + filter paper	RmT	R-FD, LN, SG. U-Oil
Trichothecium roseum	MA PDA	RmT	R-FD, LN. U-Oil
Ustilago maydis	MA PDA	RmT	R-LN, U-Oil
Verticillium lateritium	MA PDA	RmT	R-FD, LN, SG. U-Oil
Wallemia sebi	M_{20} M_{40} M_{60}	RmT	R-FD, LN. U-SG, Oil
Zygorhynchus moelleri	MA PDA	RmT	R-FD, LN. U-Oil

Key
Media abbreviations as in Appendix 1
RmT, Room temperature
R, Recommended preservation technique
U, Useful preservation technique
FD, Freeze drying
LN, Liquid nitrogen storage
SG, Silica gel storage
H_2O, Water storage
DF, Deep freeze storage
BL, Black light or Near ultra violet light (see Appendix 4)
dw, distilled water
Daylight recommended unless otherwise marked
W, Wheat straw

Appendix 3

Induction of sporulation in fungi by the use of near ultra-violet light

Light can influence many aspects of growth, development, reproduction and behaviour of fungi (Leach, 1971). The wavelenths that appear to be most effective in inducing sporulation are mainly in the near ultra-violet region (300–380 nm) of the spectrum as opposed to the ultra-violet range (200–300 nm) which can be lethal or mutagenic. In many instances the gross morphology of the colony, spore morphology or pigmentation may be affected by this range though the changes are not sufficient to intefere with their identification (Anon. 1968).

(i) Method

The fungi are grown in pyrex or clear plastic Petri dishes or bottles (Onions, 1969) for 3–4 days before irradiation and the edges sealed with clear tape to prevent rapid drying. They are illuminated on light benches which have three 122 cm fluorescent lamp holders 13 cm apart. A black light tube (Philips TL 40 W/08) is held in the centre holder and a cool white tube (Philips MCFE 40 W/33) is placed on either side. They are controlled by a time switch which is set at a 12 hour on/off cycle. The Petri dishes or bottles are supported on a shelf 32 cm below the light source and illuminated until sporulation is induced.

(ii) Light source

Cool white daylight fluorescent lamps emit a significant amount of near ultra-violet radiation and can be used if the black light tubes are unavailable. Incandescent lamps should not be used.

(iii) Considerations

Irradiation at 21°C has proved successful and at CMI the stimulation is carried out at room temperature. The illumination cycle can be controlled to suit the fungi to be stimulated. A period of darkness can be included for those fungi that require it. It is not advantageous to irradiate a culture until it has had time to grow. The media employed can have an effect on the stimulation and weak media should be used (Leach, 1962). Rich media such as potato dextrose agar (PDA) or Malt agar (MA) often give excessive growth of mycelium. *Diaporthe phaseolorum* var. *batatis* when growing on glucose casein hydrolysate medium and stimulated by near ultra-violet light produces many perithecia and ascospores. However, when grown on malt or a potato glucose medium the number of perithecia is reduced (Timnick, Lilly & Barnett, 1951).

Pigmented dematiaceous hyphomycetes, Coelomycetes and Ascomycetes respond favourably to 'black light' treatment.

Appendix 4

Prevention of mite infestations

Fungal cultures are susceptible to infestation with mites. These small animals, commonly of the genera *Tyroglyphus* and *Tarsonemus*, occur naturally in soil and on almost any organic material. They can be brought into the laboratory on fresh plant material, decaying mouldy products, on shoes, on the bodies of flying insects or even in cultures received from other laboratories. Outbreaks appear to be particularly frequent in new laboratory buildings, where new apparatus, equipment, packaging and wood appear to carry infestation, while the damp conditions of drying walls and fabric give the high humidity particularly favourable to their growth.

The damage mites cause is twofold. First they eat the cultures and a heavy infestation can completely strip a culture. Second they carry fungal spores and bacteria on and in their bodies and as they move from one culture to another the cultures can be cross inoculated and heavily infected with other fungi and bacteria.

The common fungal mites are about 0.25 mm in length. They can be seen by the naked eye only as a tiny white dot almost at the limit of vision, so infestation can easily go undetected. Given favourable conditions of high humidity and temperature they breed rapidly and spread quickly. Many cultures can be infested before they are noticed. Infested cultures have a deteriorated look and this is often the first indication of their presence.

General hygiene and preventative precautions are better than having to control an outbreak. All incoming material should be examined when it enters the laboratory and a separate room for checking and processing dirty material is desirable. The sealing of incoming cultures, storage in a refrigerator or some form of screening and quarantine system can be helpful, as it is possible for cultures with only a light infestation at the time of receipt to develop a heavy infestation later.

Methods of control used by different workers are various and a combination of precautions seems the answer. These can be classed in several categories: 1) Hygiene; 2) Fumigation; 3) Mechanical and chemical barriers; and 4) Protected storage.

(1) Hygiene

Hygiene coupled with quarantine procedures is perhaps the best protection. All work surfaces must be kept clean and cultures protected from aerial and dust contamination for example by storage on protected shelves.

The work benches and cupboards should be regularly washed with an acaricide, especially as soon as infestation is suspected. The procedure at the CMI is to wash down with an acaricide which is left for sufficient time to have effect and then cleaned off with alcohol. The benches are then repolished if desired. The acaricides used are Kelthane (Murphy Chemical Co., 9), Tedion V-18 (Midox Ltd., 10), Chlorocide (Boots Farm Sales Ltd., 11) and Actellic, an acaricide developed by ICI (15) for agricultural use. None of these has been found to be noticeably fungicidal. The acaricides cannot be allowed to remain as they are frequently toxic or skin irritant. Plastic gloves should be worn when handling them. As mites appear to become resistant to some chemicals the acaricide should be changed from time to time. When mites are found the affected cultures should be removed immediately and if possible sterilized and all cultures in the immediate area checked and isolated from the rest.

(2) Fumigation

Various chemicals are used as acaricides. They are placed in the boxes or cupboards in which the fungi are stored, either as a short treatment or for permanent storage. Perhaps the most used of these are camphor and paradichlorobenzene (PDB) (British Drug Houses Ltd., 12). Unfortunately PDB seems to have some effect on the fungi, which at least may produce abnormal growth.

Tractor vapourising oil (TVO) was formerly used at the CMI as a deterrent. It did not kill mites, but discouraged their entry into cupboards. Unfortunately TVO is now so highly purified that the deterrent factor has been removed. Smith (1967) recommended the addition of chemicals to the fungal cultures. He added Cypro and Kelthane as two drops on the bottom of the culture plug and found that this killed mites and had a low fungicidal effect. 0.05 g of PDB added as crystals killed mites but tended to affect the growth of the fungi.

(3) Mechanical and chemical barriers

Many physical methods of prevention of infestation with and spread of mites have been tried.

The culture bottles, tubes or plates can be stood on a platform surrounded by water or oil, or on a surface inside a barrier of petroleum jelly or other sticky material. These methods may be a protection from crawling mites, but do not protect from mites carried by insects or on the hands and cloths of laboratory workers.

Snyder & Hansen (1946) described a method in which culture bottles or tubes are sealed with cigarette paper. This is particularly useful for cotton wool plugged tubes as the plug can be pushed down and the tube sealed above the plug, without fear of contamination. The cigarette paper can be sterilized with propylene oxide and stuck to the rim of the culture container using a copper sulphate gelatine glue*. The pores of the paper are of a size to allow free passage of air but are too small to allow mites to pass. Care must be taken to see that a good seal is made, and that the paper is not damaged through handling. Once this method has been introduced as a routine it does not cause much trouble. It has the advantage that not only does it keep mites out, but it also keeps mites in—thus preventing spread of infestation. Smith (1971) recommends the use of disposable plastic bottles with plastic caps, which he claims can be screwed down excluding the mites but without killing the fungi.

A new type of screw-lid closure which includes a hole sealed with Metricel is said to give complete protection and is reusable as the lids can be autoclaved (Smith, 1978).

---

* *Copper Sulphate Gelatine Glue*

20 g gelatine
2 g copper sulphate
100 ml distilled water
20 g gelatine are dissolved in 100 ml water and then the copper sulphate added.
Stocks of this can be prepared and stored in culture bottles with the caps screwed down.

Sealing of Petri dishes and bottles etc. with sticky tape such as Sellotape or Scotch tape may reduce penetration but will not act as a complete barrier. Mites eventually find their way through cracks and wrinkles. Good tight cotton wool plugs will serve as a slight barrier and are much more effective than metal caps. A mite that has penetrated a cotton wool plug is somewhat cleaned and tends to carry rather less infection than others. Plugs treated with a drop of mercuric chloride kill mites moving in and out of tubes but the mercuric chloride can be poisonous to both fungi and laboratory workers so must be used with care. If the mercuric chloride solution contains a dye, then treated plugs will be easily distinguished.

(4) Protected storage

The various methods of long term storage of cultures used in culture collections prevent infestation and spread of mites, but are of little use for day to day growth of cultures.

Cold storage at 4–8°C definitely reduces the spread of mites, which are almost immobile at this temperature. However, on removal from the refrigerator the mites rapidly become active again. Storage of infested cultures in the deep freeze gives better control. Infested cultures on removal are still viable, whereas the mites are usually dead.

Covering cultures with mineral oil increases the period of viability of cultures and mites will not penetrate the oil.

Cultures stored in silica gel are kept in sealed tubes or in bottles with the caps screwed down so penetration cannot occur. Freeze dried ampoules being completely sealed are impermeable to mites and they cannot penetrate ampoules stored at the ultra low temperatures of liquid nitrogen.

Appendix 5

List of suppliers

1. Rejafix Ltd, Harlequin Avenue, Great West Road, Brentford, Middlesex.
2. Edwards High Vacuum, Manor Royal, Crawley, West Sussex.
3. Buck & Hickman, Sterling Industrial Estate, Rainham Road South, Dagenham, Essex.
4. T. W. Wingent, 115–150 Cambridge Road, Milton, Cambridge.
5. Adelphi (Tubes) Manufacturing Ltd., Duncan Terrace, London, N1.
6. Anchor Glass Co. Ltd, Brent Cross Works, North Circular Road, London.
7. Sterilin (Distributors) R & L Slaugher Ltd, 162 Balgores Lane, Gidea Park, Romford, Essex.
8. Union Carbide (Distributors) Jencons (Scientific) Ltd, Leighton Buzzard, Bedfordshire.
9. Murphy Chemical Co, Wheathampstead, St. Albans, Hertfordshire, England.
10. Mi-Dox Ltd, Smarden, Kent.
11. Boots Farm Sales Ltd, Nottingham, Nottinghamshire.
12. British Drug Houses Ltd, Laboratory Chemicals Division, Poole, Dorset.
13. Scott's Office Equipment Ltd, Deseronto Wharf, St. Mary's Road, Langley, Slough, Berkshire, SL3 7EW.
14. Denley Instruments Ltd, Daux Road, Billinghurst, Sussex, RH14 9SJ.
15. Agricultural Suppliers or direct from: Imperial Chemical Industries Ltd, Mond Division Sales, Star House, Clarendon Road, Watford.
16. Edme Ltd, Mistley, Manningtree, Essex.

Mention of commercial organisations or their products does not necessarily imply CMI recommendation but is included as an indication of availability.

References

Alexander, M.; Daggett, P. M.; Gherna, R.; Jong, S.; Simione, F.; Hatt, H. (1980) *American Type Culture Collection Methods I. Laboratory Manual on Preservation Freezing and Freeze-drying.* Rockville, Maryland: American Type Culture Collection.

Anon. (1968) *Mites leaflet* 4pp. Kew; Commonwealth Mycological Institute.

von Arx, J. A.; Schipper, M. A. A. (1978) The CBS fungus collection. *Advances in Applied Microbiology* **24**, 215–236.

Ashwood-Smith, M. J.; Grant, E. (1976) Mutation induction in bacteria by freeze drying. *Cryobiology* **13**, 206–213.

Ashwood-Smith, M. J.; Warby, C. (1971) Studies on the molecular weight and cryoprotective properties of PVP and dextran with bacteria and erythrocytes. *Cryobiology* **8**, 453–464.

Atkinson, R. G. (1953) Survival and pathogenicity of *Alternaria raphani* after 5 years in dried soil cultures. *Canadian Journal of Botany* **31**, 542–547.

Baker, P. R. W. (1955) The micro-determination of residual moisture in freeze-dried biological materials. *Journal of Hygiene* **53**, 426–435.

Boeswinkel, H. J. (1976) Storage of fungal cultures in water. *Transactions of the British Mycological Society* **66**, 183–185.

Booth, C. (1971) *The genus Fusarium.* 237 pp. Kew: Commonwealth Mycological Institute.

Buell, C. B.; Weston, W. H. (1947) Application of the mineral oil conservation method to maintaining collections of fungus cultures. *American Journal of Botany* **34**, 555–561.

Butterfield, W.; Jong, S. C.; Alexander, M. J. (1978) Polypropylene vials for preserving fungi in liquid nitrogen. *Mycologia* **70**, 1122–1124.

Calcott, P. H. (1978) *Freezing and thawing microbes.* Meadowfield Press, England.

Carmichael, J. W. (1962) Viability of mould cultures stored at −20°C. *Mycologia* **54**, 432–436.

Castellani, A. (1939) Viability of some pathogenic fungi in distilled water. *Journal of Tropical Medicine and Hygiene* **42**, 225–226.

Castellani, A. (1967) Maintenance and cultivation of common pathogenic fungi of man in sterile distilled water. Further researches. *Journal of Tropical Medicine and Hygiene* **70**, 181–184.

Cooney, D. G.; Emerson, R. (1964) *Thermophilic fungi.* xii + 188pp. San Francisco and London: W. H. Freeman.

Dade, H. A. (1960) Laboratory methods in use in the culture collection, CMI In *Herb. I. M. I. Handbook*, 78–83. Kew: Commonwealth Mycological Institute.

Diener, U. L. (1955) Sporulation in pure culture by *Stemphylium solani Phytopathology* **45**, 141–145.

Dietz, A. (1975) In *Round Table Conference on Cryogenic Preservation of Cell Cultures* (Rinfret, A. P.; La Salle, A. B. eds) 22–36. Washington, DC: National Academy of Science.

Al Doory, Y. (1968) Survival of dermatophyte cultures maintained on hair. *Mycologia* **60**, 720–723.

Elliott, T. J. (1976) Alternative ampoule for storing fungal cultures in liquid nitrogen. *Transactions of the British Mycological Society* **67**, 545–546.

Ellis, J. J. (1979) Preserving fungus strains in sterile water. *Mycologia* **71**, 1072–1075.

Ellis, J. J.; Roberson, J. A. (1968) Viability of fungus cultures preserved by lyophilization. *Mycologia* **60**, 399–405.

Figueiredo, M. B. (1967) Estudos sôbre a splicação de método de Castellani para conservação de fungos patógenos em plantas. *Biológico* **33**, 9–13.

Figueiredo, M. B.; Pimentel, C. P. V. (1975) Métodos utilizados para conservação de fungos na micoteca da Seção de Micologia Fitopatológica do Instituto Biológico. *Summa Phytopathologica* **1**, 299–302.

Galindo, J.; Gallegly, M. E. (1960) The nature of sexuality in *Phytophthora infestans. Phytopathology* **50**, 123–128.

Goldie-Smith, E. K. (1956) Maintenance of stock cultures of aquatic fungi. *Journal of the Elisha Mitchell Scientific Society* **72**, 158–166.

Gordon, W. L. (1952) The occurrence of *Fusarium* species in Canada. *Canadian Journal of Botany* **30**, 209–251.

Hawksworth, D. L.; Sutton, B. C.; Ainsworth, G. C. (1983) *Ainsworth & Bisby's Dictionary of the fungi.* 7th edition, 457 pp. Kew: Commonwealth Mycological Institute.

Heckly, R. J. (1978) Preservation of microorganisms. *Advances in Applied Microbiology* **24**, 1–53.

Hwang, S.-W. (1960) Effects of ultra-low temperature on the viability of selected fungus strains. *Mycologia* **52**, 527–529.

Hwang S.-W. (1966) Long term preservation of fungal cultures with liquid nitrogen refrigeration. *Applied Microbiology* **14**, 784–788.

Hwang, S.-W. (1968) Investigation of ultra low temperature for fungal cultures I. An evaluation of liquid nitrogen storage for preservation of selected fungal cultures. *Mycologia* **60**, 613–621.

Hwang, S.-W.; Howells, A. (1968) Investigation of ultra low temperature for fungal cultures. II. Cryoprotection afforded by glycerol and dimethyl sulphoxide to 8 selected fungal cultures. *Mycologia* **60**, 622–626.

Hwang, S.-W.; Kwolek, W. F.; Haynes, W. C. (1976) Investigation of ultra low temperature for fungal cultures. III. Viability and growth rate of mycelial cultures following cryogenic storage. *Mycologia* **68**, 377–387.

Leach, C. M. (1962) The quantitative and qualitative relationship of monochromatic radiation to the induction of reproduction in *Aschochyta pisi. Canadian Journal of Botany* **40**, 1577–1602.

Leach, C. M. (1971) A practical guide to the effects of visible and ultraviolet light on fungi. In *Methods in Microbiology,* **4** (Booth, C. ed.), 609–664. London and New York: Academic Press.

Ogata, W. N. (1962) Preservation of *Neurospora* stock cultures with anhydrous silica gel. *Neurospora Newsletter* **1**, 13.

Onions, A. H. S. (1969) Disposable polystyrene containers for improved sporulation of stock cultures. *Transactions of the British Mycological Society* **52**, 173–174.

Onions, A. H. S. (1971) Preservation of fungi. In *Methods in Microbiology,* **4**, (Booth, C., ed.): 113–151. London and New York; Academic Press.

Onions, A. H. S. (1977) Storage of fungi by mineral oil and silica gel for use in the collection with limited resources. In *Proceedings of the Second International Conference on Culture Collections.* Brisbane; World Federation of Culture Collections.

Onions, A. H. S. (1983) Preservation of fungi. In *The filamentous fungi,* **4**, (Smith, J. E.; Berry, D. R., eds): 373–390. London: Edward Arnold.

Perkins, D. D. (1962) Preservation of *Neurospora* stock cultures in anhydrous silica gel. *Canadian Journal of Microbiology* **8**, 591–594.

Pitt, J. I. (1973) An appraisal of identification methods for *Penicillium* species: Novel taxonomic criteria based on temperature and water relations. *Mycologia* **65**, 1135–1157.
Pitt, J. I. (1979) [publ. 1980] The genus *Pencillium* and its telemorphic states *Eupenicillium* and *Talaromyces*. 634 pp. London, New York, Toronto, Sydney, San Francisco: Academic Press.
Prescott, J. M.; Kernkamp, M. F. (1971) Genetic stability of *Puccinia graminis* f.sp. *tritici* in cryogenic storage. *Plant Disease Reporter* **55**, 695–696.
Raper, K. B.; Alexander, D. F. (1945) Preservation of molds by the lyophil process. *Mycologia* **37**, 499–525.
Raper, K. B.; Thom, C. (1949) *A manual of the Penicillia.* ix + 875pp. Baltimore: Williams and Wilkins.
Reinecke, P.; Fokkema, N. J. (1979) *Pseudocercosporella herpotrichoides:* Storage and mass production of conidia. *Transactions of the British Mycological Society* **72**, 329–331.
Reischer, H. S. (1949) Preservation of Saprolegniaceae by the mineral oil method. *Mycologia* **41**, 177–179.
Rey, L. R. (1977) Glimpses into fundamental aspects of freeze drying. In *Developments in biological standardisation. International symposium on freeze drying of biological products* (Cabasso, V. J.; Regamey, R. H.; eds): 19–27. Basel; S. Karger.
Shearer, B. L.; Zeyen, R. J.; Ooka, J. J. (1974) Storage and behaviour in soil of *Septoria* species isolated from cereals. *Phytopathology* **64**, 163–167.
Simione, F. P.; Daggett, P. M.; McGrath, M. S.; Alexander, M. T. (1977) The use of plastic ampoules for freeze preservation of microorganisms. *Cryobiology* **14**, 500–502.
Smith, D. (1982) Liquid nitrogen storage of fungi. *Transactions of the British Mycological Society* **79**, 415–421.
Smith, D. (1983a) A two stage centrifugal freeze drying method for the preservation of fungi. *Transactions of the British Mycological Society* **80**, 333–337.
Smith, D. (1983b) Cryoprotectants and the cryopreservation of fungi. *Transactions of the British Mycological Society* **80**, 360–363.
Smith, D.; Onions, A. H. S. (1983) The comparison of some preservation techniques for fungi. *Transactions of the British Mycological Society*, **81**, 535–540.
Smith, R. S. (1967) Control of tarsonemid mites by cigarette paper barriers. *Mycologia* **59**, 600–609.
Smith, R. S. (1971) Maintenance of fungal cultures in pre-sterilized disposable screwcap plastic tubes. *Mycologia* **63**, 1218–1221.
Smith, R. S. (1978) A new lid closure for fungal culture vessels giving complete protection against mite infestation and microbiological contamination. *Mycologia* **70**, 499–508.
Snyder, W. C.; Hansen, H. N. (1946) Control of culture mites by cigarette paper barriers. *Mycologia* **38**, 455–462.
Staffeldt, E. E.; Sharp, E. L. (1954) Modified lyophil method for preservation of *Pythium* species. *Phytopathology* **44**, 213–214.
Timnick, M. B.; Lilly, V. G.; Barnett, H. L. (1951) Factors affecting sporulation of *Diaporthe phaseolorum* var. *batatis* from soybean. *Phytopathology* **41**, 327–336.
Tuite, J. (1968) Liquid nitrogen storage of fungi sealed in polyester film. *Mycologia* **60**, 591–594.

Index

Acaricides 43
Actellic 43
Actinomycetes
Aeration 3
Ampoules 13, 15, 16, 17, 18, 19, 21, 22, 23
Argon 14
Ascomycetes 8, 25, 41
Ascomycotina 11, 18, 23
Ascospores 41
Aspergilli 1, 5, 7, 11
a/w (water activity) 3
Basidiomycetes 7, 10, 25, 30
Basidiomycotina 11, 18, 23, 29
Black light (near ultra violet light) 3, 5, 39, 41
Borosilicate glass ampoules 22
Camphor 44
Centrifugal freeze drying 16, 17, 18, 19, 30
Chlorocide 43
Chytridiomycetes 18, 23
Coelomycetes 18, 23, 25, 41
Cold hardening 22
Cold (cool) storage 5, 45
Cooling
  by evaporation 16
  rate 13, 20, 22
Cryoprotectant 21
  dimethyl sulphoxide 21
  glycerol 21, 22
  polyvinyl pyrrolidone (PVP) 21
Daylight 39
Deep freeze 6, 27, 28, 29, 30, 39, 45
Dematiaceous hyphomycetes 41
Dermatophytes 2
Desiccation 11
Deuteromycotina 11, 18, 23, 29
Dimethyl sulphoxide 21
Dormancy 8, 11, 20
Entomophthorales 29, 30
Evaporative freezing 13
Freeze drying 12–20, 27, 28, 29, 30, 37, 38, 39
Frequent transfer 5, 27, 29
Fumigation
  prevention of mite infestation 43, 44
Glycerol 21, 22
Growth of cultures 1–4
Humidity 5, 6
Hyphochytriomycetes 23
Hyphomycetes 11, 18, 23, 41
Kelthane 43
Laboulbeniales 29
Light 7
  daylight 39
  sunlight 37, 38
Liquid nitrogen storage 20–25, 27, 28, 29, 30, 37, 38, 39
Loam 8, 9
Lyophilisation 12
Mastigomycotina 11, 18, 23, 25, 29, 30
Media 1, 2
  cornmeal (maize) agar (CMA) 1, 31, 37, 38
  Czapek (dox) agar (Cz) 1, 31, 37, 38
Czapek yeast autolysate (CYA) 1, 32
  egg yolk medium 1, 32
  glucose hydrolysate medium 41
  hemp seed 37, 39
  malt czapek agar (MCz) 32, 37, 38
  malt extract agar (MEA) 1, 32
  malt extract agar (MA) 1, 33, 37, 38, 39
  malt extract + 20% sucrose (M_{20}) 33, 37, 39
  malt extract + 40% sucrose (M_{40}) 33, 37, 39
  malt extract + 60% sucrose (M_{60}) 33, 39
  oat agar (OA) 1, 33, 37, 38, 39
  onion seed 39
  potato carrot agar (PCA) 1, 33, 37, 38, 39
  potato dextrose agar (PDA) 1, 33, 37, 38, 39, 41
  potato glucose medium 41
  potato sucrose agar (PSA) 1, 34, 38
  rabbit dung agar (RDA) 34, 37, 38, 39
  Sabouraud's medium—Sabouraud's conservation media (SCA) 34
  soil extract agar (SEA) 34, 38
  starch agar (SA) 35, 38
  steep agar 1, 35
  tap water agar 1, 35, 37, 38
  V8 juice agar 35, 37
  yeast peptone soluble starch agar (YPSS) 36, 37, 38, 39
Mercuric chloride 45
Mesophile, mesophilic 2
Mineral oil storage 6–8, 27, 28, 29, 30, 37, 38, 39
Mites 5, 43–45
Mucorales 1
Mutagenic 41
Mutation 13
Mycelial 8, 22
  basiodiomycetes 7
Mycelium 10
Near ultra violet light (black light) 39, 41
Nitrogen 18
  see liquid nitrogen
Oil storage
  see mineral oil storage
Oomycetes 11, 12, 18, 23, 29, 30
Osmophile, osmophilic 3
Oxygen tension 6
Paradichlorobenzene (PDB) 44
Pathogenicity 9
Penicillia 1, 5, 7, 11
Perithecia 41
pH 1, 4
Phycomycetes 10
Plant Pathogens 10
Plastic
  ampoules 21
  film 21
  straws 21
Polyvinyl pyrrolidone (PVP) 21
Psycrophiles 2
Psycrophilic 2
Psycrotolerant 2
Refrigerator 5, 8, 9, 21, 22, 23, 29
Regular transfer
  see Frequent transfer
Rehydration 14
Residual moisture 13, 17
Room temperature storage 5, 17
Rusts 29
Saprolegniaceae 7
Sclerotia 17, 30
Shelf (freeze) drier 17, 20
Silica gel storage 11–12, 28, 29, 30, 39
Smuts 29
Soil storage 8–9, 28, 29, 30, 39
Sporulation 41
Sunlight 37, 38
Suspending media
  glucose 13
  inositol 16
  peptone 13, 16
  serum 13, 16
  skimmed milk 13
  sucrose 13
Tarsonemus 43
Tedion 43
Temperature 1, 21, 37, 38, 39
Thawing 22, 24
Thermophile 2
Thermophilic 2
Thermotolerant 2
Tractor vapourising oil (TVO) 44
Tyroglyphus 43
Ultra-low temperature 20
  see liquid nitrogen storage
Ultra-violet light 13, 39, 41
Uredinales 29
Uredospores 38
Ustilaginales 29, 30
Vials 13, 14, 15, 17
Water
  activity a/w 3
  moulds 7
  storage 9–10, 28, 29, 37, 38, 39
Xerophile 3
Zygomycetes 11, 18, 23
Zygomycotina 11, 18, 23, 29